THE HIDDEN LANGUAGE OF CATS

THE HIDDEN LANGUAGE OF CATS

HOW THEY HAVE US AT MEOW

Sarah Brown

DUTTON

DUTTON

An imprint of Penguin Random House LLC
penguinrandomhouse.com

Illustrations by Hettie Brown

LIBRARY OF CONGRESS CATALOGING-IN-PUBLICATION DATA

Names: Brown, Sarah L., author.
Title: The hidden language of cats: how they have us at meow / Sarah Brown.
Description: New York : Dutton, Penguin Random House, [2023] |
Includes bibliographical references and index.
Identifiers: LCCN 2023009025 |
ISBN 9780593186411 (hardcover) | ISBN 9780593186435 (ebook)
Subjects: LCSH: Cats—Behavior. | Human-animal communication.
Classification: LCC SF446.5 .B78 2023 | DDC 636.8—dc23/eng/20230711
LC record available at https://lccn.loc.gov/2023009025

Printed in the United States of America
1st Printing

Book design by Nancy Resnick

For the cats

CONTENTS

INTRODUCTION

It was the late 1980s when, as a young scientist, I embarked on an exciting adventure. My mission was to study the behavior of domestic cats, starting with how those living in neutered groups interact with one another, and how cats communicate with people. More than thirty years and countless cats later, I am still on that adventure.

It turns out that studying cat behavior involves a delicate and sometimes challenging balance between rigorous science and unbridled delight at your subjects. In his 1911 book, *Animal Intelligence*, Edward Thorndike, one of the earliest pioneers of experimental psychology, referred rather disparagingly to "the well-nigh universal tendencies in human nature to find the marvelous wherever it can." He felt that this habit inevitably led to somewhat biased judgments in both choosing what to study and interpreting the results. In other words, a serious animal behavior scientist should be as objective as possible, avoiding at all costs the temptation to sing the praises of their subjects.

At the beginning of my doctoral degree, as I set about my research, I was desperately keen to discover new facts about cats. With Thorndike's words echoing in my mind, however, I knew my studies had to be carefully planned and analyzed to make their mark as "proper" science. I duly gathered my data, made my deductions, and gained my doctorate in a sensible, scientific manner. And yet, from my very first day and the very first cats, I found myself marveling over and over at these enigmatic creatures—at their adaptability, ingenuity, and resilience.

This book looks at how domestic cats, descended from solitary North African wildcats, have managed to set up home with devoted owners all over the world. In the USA alone, over forty-five million households are now home to at least one cat. How did they do it? How did those wildcats of old creep into our homes and our hearts and convince us that we should keep them warm, fed, and pampered? Quite simply, they learned to talk to us. They also learned to talk to each other, a fact rarely acknowledged when cats are compared with dogs in the endless competition to be our best friend. Dogs are descended from wolves, a social species from which they inherited a finely honed repertoire of interactive behavior patterns—a comprehensive manual of how to speak to others. Cats, on the other hand, inherited very few social skills from their poker-faced wildcat ancestors, who rarely came face-to-face with one another. They have had a far greater social journey to make than our humble hounds.

Dipping into my own and other scientists' discoveries, we will learn how cats supplemented their original scent-based language with new signals and sounds designed for life alongside humans and other cats. Despite this monumental effort on their part to communicate more effectively, how much do we really understand of their language and vice versa? How do cats perceive us? Do they think of us as their "owners," or more like large two-legged cats with a poor sense of smell? *The Hidden Language of Cats* explores the science that answers these and many more questions—and features some of the marvelous cats that have helped along the way.

The Marvelous Cats

My battered old student car sputtered its way around the bend and up the last stretch of hill. As it reached the brow, a huge, imposing building crept gradually into sight: a sprawling red-brick Victorian hospital in the middle of nowhere, like something from a gothic novel. I tentatively drove in and surveyed the scene. The hospital had opened to patients in 1852, with the somewhat dubious title of "county lunatic asylum." By the time I paid my first visit, over 130 years later, it was still very much a working institution, although now more appropriately described as a psychiatric hospital. I was more interested in the activity outside its doors though—I had come in search of cats to watch.

It was with some relief that I saw the wonderfully welcoming and helpful head porter, John, who gave me a guided tour of the grounds. He explained how the hospital's infamous cat population was a mixture of completely unapproachable feral individuals and a bunch of more friendly cats that would come up and say hello to you. This mixed population was a result of the hospital being, for many years and by virtue of its remoteness, a notorious local dumping ground for unwanted pet cats. The original collection of dumpees had reproduced among themselves, and in the absence of regular human contact, the subsequent generations of kittens became progressively more feral and reserved. Meanwhile, there were constant new recruits to the population, courtesy of dissatisfied owners arriving in cars under cover of darkness and depositing their poor, unsuspecting pet cats. These newbies were usually the friendly ones, fresh out of human homes and still up for a chat with passing people.

John wasn't sure exactly how many years cats had been there, but there are records as far back as the 1960s of the nurses feeding them on their breaks between shifts. Regular attempts were made to neuter the cats, but with the constant influx of new arrivals, it was a never-ending battle.

We saw a lot of cats that day. As John had predicted, some were well fed, content, and obviously used to people, and they would approach us as we passed through the areas in which they lay sunning themselves. Others you could be forgiven for completely missing; a quick darting movement out of the corner of your eye and they were gone.

The grounds of the hospital were vast, and due to the design of the buildings, the various wings branched out in different directions, forming natural courtyards. This allowed groups of cats to separate themselves from one another, creating smaller subpopulations. The buildings had an extensive system of basements, through which ran large heated pipes—a kind of old-fashioned underfloor heating system. Ventilation shafts ran from these warm basements to openings in the brick walls of the courtyards—these were premium resting spots for cats. As we walked, I could make out raggedy furry faces peering out of each opening, along with more cats dotted around the courtyards, hoping for a turn. For some time, I had been searching for a small stable group of feral cats to watch for my studies of cat-to-cat interactions. This large hospital population contained a number of such groups, and as I drove away that day, I decided this was the ideal place to begin my research.

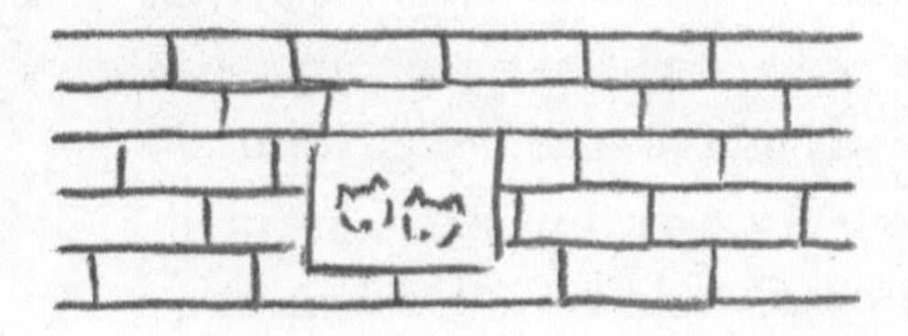

Faced with a seemingly endless sea of cats, my early days looking for the right subgroup to study were daunting, to say the least. I began making patrols of the grounds and recording, on printed cat outlines, the details of all the cats I saw. I would annotate the blank images with cats' markings, including views from the left side, right side, and face-on, so I had a sort of feline mug shot for each individual. After a few weeks, I began to get a handle on the population—which cats hung out where, which ones roamed, and which ones tended to stay in the same places. I made notes on them too—looking back at these makes for some interesting reading. While my "Boiler room cat" entry reads "black and white, red collar, friendly," the "Electrician's cat" is described as "black, large with a white collar, HATES WOMEN." I seem to have only one recorded observation of the electrician's apparently misogynistic cat.

As a pattern emerged of the distribution of cats around the hospital, one particular group that frequented a small courtyard caught my attention. Food was provided in this courtyard around the same time each day—leftovers from the wards were put outside for the eagerly waiting cats to eat. This regular provision of food and the availability of plenty of resting places meant that the cats stayed around this one specific courtyard, creating what appeared to be a relatively stable group. Of the five core members of the group, one, whom I named Frank, would regularly leave the yard to explore farther afield, always returning to eat with the other four. I found I could rely on them being in this location each day, which, along with the ease of observing them there from a discreet distance, meant they were an ideal group on which to base my more detailed observations of cat social interactions. So Betty, Tabitha, Nell, Toby, and Frank became my first study group—a group you'll get to know more throughout the upcoming chapters.

Alongside my studies of the hospital cats, I began to look for a second feral colony to watch. I was working at that time as a research assistant at the Anthrozoology Institute based at the University of Southampton, England. One day we received a call about a group of cats living beneath the buildings of a local school. The headmaster and school governors wanted to be rid of them, and so a rescue mission was organized in conjunction with a local animal shelter that was well practiced at trapping such unapproachable feral cats. We arrived one evening in various cars piled high with humane cat traps and cans of tuna. A couple of hours later we left, traps baited and primed, to wait and see if any of the cats would get hungry enough overnight to venture into them. The next few days revealed much about the personalities of the individual colony members. Some had been easy to tempt into the traps and resisted little when carted off to the rescue shelter and vet to be treated for worms and fleas, vaccinated, neutered, and, of course, fed and watered. Others were more of a challenge and took a little more persuading. And then there was Big Ginger, as he became known. That large, battered ginger face peered from inaccessible nooks and crannies for many days before he could resist temptation no longer and succumbed to the lure of sardines one dark night. Eventually we had them all.

Although they would probably have protested otherwise, Ginger and his gang were lucky as feral cats go. Some of the females were pregnant and so gave birth to and raised their kittens in the safe and warm environment of the rescue center. All the kittens found homes—they were still young enough to socialize with people. The adult cats, however, were too set in their ways to begin a life with humans. We found them a new outdoor home on an old farm that was being used as a tree nursery. Here we were permitted to set up a shed as a base and were allowed daily access to feed and observe the cats. My colleagues helped with the feeding on the days I couldn't make it, but whenever I could, I pitched up and watched the colony for a few hours, doling out cans of food just before I left. I gave the cats names: Sid, Blackcap, Smudge, Penny, Daisy, Dusty, Gertie, Honey, Ghost, Becky, and, of course, Big Ginger.

The hospital and farm colonies became my life for the next couple of years. Largely aloof and under-socialized, the cats mostly kept their distance from me and barely acknowledged my presence. But this was what I wanted—to watch them doing what cats do around other cats.

As Ernest Hemingway wrote, "One cat just leads to another." And sure enough, other cats appear in this book, characters I met over the years who each helped me learn a little more of what it must be like for cats living with cats, and for cats living with humans. I got to know many wonderful owners and their pet cats when working as a behavior counselor—they taught me so much about the different relationships people have with their cats. Mrs. Jones and her lovely Cecil were one such pair, featured in chapter two. My own pet cats, companions with whom I have shared different parts of my life, also pop up from time to time—Bootsy, Smudge, Tigger, and Charlie. These are probably the cats I know best—when living under the same roof as a cat it becomes much easier to learn their language, rather like moving to a foreign country and immersing yourself in its language and culture.

Dotted here and there throughout the chapters are some of the unforgettable cats I met while working in a rescue shelter—Ginny, Mimi, Pebbles, and Minnie. Plus Sheba, who I fostered in my home while she raised her kittens. My studies of the hospital and farm cats as a doctoral student had opened my eyes to the struggles of cats born in unsafe environments and unable to receive proper food and veterinary care. After working briefly with a rescue organization when relocating the farm cats, I realized if I was to properly experience cats and their myriad lifestyles, I needed to spend more time in the rescue environment. I promised myself I would do this one day, although it was not until around thirty years later that I finally stepped over the threshold of my local cat shelter and immersed myself in that world.

Working in a cat rescue shelter was a complete revelation. The feeling of slight dread arriving in the early morning and discovering a large cardboard box abandoned on the doorstep of the shelter. Opening it to reveal inside a grumpy and battered old tomcat with tattered ears or, as springtime approached, a litter of scrawny flea-ridden kittens. These rescue cats made lasting impressions on me and reinforced my wonder and admiration for this most resilient of species and its ability to morph from street-living stray to affectionate lap cat in the space of a week.

We also meet, briefly, a couple of dogs, Alfie and Reggie, who have lived alongside my cats Bootsy and Smudge at different times. They provide classic examples of how both cats and dogs learn to rub along (quite literally) with one another, as well as with members of their own species and with people—yet another language for them all to learn.

The imposing hospital where I'd studied my first group of cats closed down in 1996, a few years after I had completed my doctorate. It was eventually converted into beautiful, high-ceilinged apartments. What became of all the cats remains a mystery—I like to think they found somewhere new to live and glean handouts locally. The farm cats, too, were later moved to a new farm. They were harder to watch in their new location, but my studies were largely finished by then, and they got to live out the rest of their lives well provisioned, peaceful, and content. As indeed should all cats.

CHAPTER 1

WILDCATS AND WITCHES

Cat said, "I am not a friend, and I am not a servant. I am the Cat who walks by himself, and I wish to come into your cave."

—Rudyard Kipling, *The Cat That Walked by Himself*

I stood in the corridor outside the walk-in cat pen at the rescue center and watched through the wire mesh door. The center manager, Ann, already inside the pen, approached the big, glaring ginger cat who was crouched against a side wall, eyes like saucers, fur fluffed to the maximum, hissing and growling in a truly terrifying manner. Undeterred, Ann wielded her syringe of vaccine and, with fearless skill and dexterity, quickly jabbed him with the needle. Big Ginger, as we later named him, launched himself. Not at Ann, but up the wall, across the ceiling, down the other side, and into a box to hide in the blink of an eye. Retracing his route with my eyes, I asked Ann, "Did he actually just run over the ceiling?" She smiled. "The ferals often do that." Rookie postgrad that I was, I had to confess to her this was my first proper experience with feral cats, which are unsocialized domestic cats that have reverted to a semi-wild existence. People had laughed when I said I was going to study domestic cat behavior for my doctorate. "Domestic cats? Aren't they a bit boring? Don't you want to go abroad and study big wildcats somewhere?" I figured this one was probably wild enough for me.

While Big Ginger and the other cats from his colony were being looked after at the rescue center, my colleagues and I visited the cats' future home, a farm. We put up a shed as a base from which to feed them and, lining the shelves inside with beds, cut a cat-size hole in the door so they could use it for shelter too. A few months later, to offer them some extra cover, next to the shed we built a square wooden structure with a hinged lid containing four compartments, separated from one another internally and each with its own entrance hole. We grandly named it the "catterama."

The day after the cats were released at the farm, I stood hopefully by the shed, a tin of cat food in my hand, and surveyed the land around. There was not a cat in sight. Occasionally a flash of black and white or ginger would catch my eye and then be gone. At one point I could just about make out two little eyes reflecting back at me from the darkness of the bushes nearby. Hmm, so much for a "study" colony, I thought—would any of them ever venture out into the open again?

As I embarked on my studies of cats and their communication, words like *tame*, *feral*, *domestic*, *socialized*, and *wildcat* floated through the literature in a bewildering fashion. So much to disentangle. What did they all mean? Can you tame a wildcat? What, really, is a domesticated animal? And is a feral cat still a domestic cat? Slowly, as I learned more about Big Ginger, his colony mates, and their ancestors, I began to find answers to my questions. I realized that when looking at how my colony cats communicated, it was important to consider the history of cats and how they have adapted and changed. For example, the life of a wildcat is so different from that of a domestic cat, there surely had to be differences in their language too.

Domestication

Is the "domestic" cat actually domesticated? It is a question that's been asked time and time again, the cause of endless debates and raised fur among cat-loving and cat-hating communities around the world. Looking for an answer requires some consideration of the difference between a tame animal and a domesticated one and where the modern-day cat fits in.

Taming describes the process whereby an animal becomes biddable and often friendly toward the handler over the course of its lifetime. It applies to a single animal, not a population or species. Wild individuals of many species are tamed by people and have been for millennia.

Domestication, on the other hand, is a much longer process that involves genetic change in a whole population over time. Humans have been trying to domesticate animals, to adapt them to living with us under our terms, for thousands of years. While we have succeeded with some—like dogs—for other species it has proved an impossible challenge. Often the best result we can achieve is taming, and with many animals, even that option remains elusive.

The challenge is that for domestication to occur, a species needs certain qualities. The first, and most important, is approachability and the potential to be handled by humans—that is, they must possess the capacity to become tame. For tameness to develop into domestication, the general rule of thumb is that the animals must have the ability to live in social groups or herds controlled by a leader (and be accepting of humans in this role). They must also be flexible with their diet, eating whatever we have available to feed them. In particular, for domestication to progress, animals must be able to breed in captivity, again under the control of humans who select individuals that possess the most favorable traits. All in all, a big ask for many species of animal—not least the cat.

How do we tell if a species is domesticated? In 1868, Charles Darwin noted, with some intrigue, how domesticated mammals have certain behavioral and physical characteristics in common with one another compared with their wild ancestors. As well as the expected increase in friendliness toward people, there were odd things such as smaller brains and coat color variations. Ninety years later, in a remote research station in Siberia, what is probably the most famous ongoing domestication study in history began. Russian scientists Dmitri Belyaev, Lyudmila Trut, and their team re-created the domestication process starting with a captive population of silver foxes that had originally been reared for their luxurious fur. Although the foxes all appeared very wild, there was some natural variation in their behavior toward people. Belyaev selected those that were least reactive to approach by humans and bred from them. He then chose the tamest offspring of these matings and bred from them and so on until, after only ten generations, he had a small population of friendly, waggy-tailed, vocal, and interactive foxes. As more generations were bred, the foxes started to display physical changes, too, such as spotted coats, floppy ears, and shorter, curlier tails. Amazingly, these traits appeared simply as a side effect of selection for tameness.

Domestication syndrome, as it is now described, refers to an array of both physical and physiological traits exhibited by species that have undergone domestication. The list has grown over the years as Belyaev's fox study and others have identified additional traits, including smaller teeth, a tendency toward more juvenile facial features and behavior, reduced stress hormone levels, and a change in the reproductive cycle.

Most domestic animals exhibit a selection of these changes but rarely all of them, their expression varying among species. With so much variability, some scientists have begun to question whether domestication "syndrome" as such exists. Even Belyaev's studies have come under deeper scrutiny with the discovery that the original foxes on his farm came from fur farms in Canada and may therefore have already undergone some previous selection for handleability. While the debate about an overall syndrome continues, there seems little doubt that domestication does bring about some physical as well as genetic changes in many species compared with their wild ancestors.

Interestingly, these types of changes have also been observed in contemporary populations of certain undomesticated species. With more and more species adapting to thrive near people, some are starting to exhibit traits similar to those of domesticated species. In the UK, for example, red foxes have become increasingly present in urban areas where they show reduced fear of people. Some of these urban foxes have been found to have shorter and wider snouts and narrower brain cases compared with rural foxes, physical changes that resemble those associated with domestication in other species.

"Domesticated" cats show a few physical features that distinguish them a little, but not a whole lot, from their wildcat ancestors. Their legs are a bit shorter, their brains slightly smaller, and they have longer intestines. Domesticated cats' coats vary in color and pattern, too, compared with the consistently striped (mackerel) tabby markings of the wildcat. Floppy ears, however, do not occur, and neither do shorter, curlier tails. That there are so few obvious physical differences between them and the wildcat has caused many to question how domesticated the cat is.

So just how qualified are cats in the domestication stakes? They certainly have the capacity to become tame. On the whole, they seem happy to eat what we feed them (apart from those who have perfected the art of fussiness)—their longer intestine compared with that of wildcats is thought to be an adaptation to feeding off human scraps. They have also adapted to living in groups, although mostly only where necessary or advantageous to them. However, the list fizzles out around there. That cats regard humans as their "leaders" seems highly questionable. And perhaps because of this there is another, much bigger gap in the cat's qualification for true domestic status. Although cats are able to reproduce in captivity, selective breeding by humans to produce those with known pedigrees is a relatively recent phenomenon, dating from around the late 1800s. The popularity of such pedigreed cats as pets has grown in recent years, yet surveys indicate that still only 4 percent of owners in the US and 8 percent in the UK acquire their cats from a specialist breeder. Most domestic cats are what are known as random bred, with mixed or unknown parentage. Some are lucky enough to live as well-cared-for pets, either permanently indoors or with outdoor access, but millions of cats worldwide have no home and live very different lives, often quite independently from humans. Today many pet house cats are neutered, itself a form of breeding control by humans, albeit preventive rather than selective in nature. However, huge numbers of pet cats remain unneutered, many of them wandering freely outdoors, and these cats, along with the millions of unowned ones, form a vast reproductively intact cat population on the lookout for mates. These cats breed indiscriminately, very much not under the control of people, although often literally right on our doorsteps. Some say this widespread lack of human influence over cats' choice of mates means that they aren't fully domesticated. As a result, cats have been variously described as semidomesticated, partially domesticated, or commensal in their unique relationship with humankind.

Socialization and Feralization

However we decide to label it, the modern-day "domestic" cat does possess a genetic predisposition for friendliness toward humans. It is only a predisposition, though, and cats aren't just magically friendly to humans from the minute they're born. Kittens must first meet humans at a very early age—between two and seven weeks old—in order to become tolerant of and friendly toward humans as adults.

Take this scenario. A friendly, socialized female pet cat, let's call her Molly, falls on hard times. Molly's owners move away and abandon her, so she resorts to a life on the street, finding food where she can. If unneutered, she may become pregnant courtesy of a wandering tomcat and give birth to kittens, tucked away wherever she can find a safe, sheltered spot. These kittens may not encounter a human in the first two months of their lives, even if Molly is still friendly toward people, because, to the best of her ability, she will hide her babies from any potential danger. If left long enough without human contact, the kittens will grow up nervous toward people and will avoid them for the rest of their lives, often hanging around human habitation to glean food but avoiding interaction. As these growing kittens

then breed with other stray cats, their offspring and successive generations become increasingly wary of people.

These are known as feral cats. They are still genetically identical to domestic cats and retain the domestic cat's ability to live in close proximity to other cats when necessary. This is usually to exploit a local concentrated abundance of food, such as handouts from a restaurant or scraps from waste bins. Groups of feral cats often become established in an area, and if allowed to reproduce, they rapidly expand in number to form larger colonies.

It's not just a one-way process, though. Molly's kittens could become quite feral within a generation through missing out on socializing with people. But as domestic cats, they still retain and genetically pass on the ability to be friendly toward humans if socialized. The progeny of these potential ferals, if introduced to people early enough, could become properly socialized and live with people as happily adjusted pet cats, just like their grandmother Molly once did.

It was in just such a colony that Big Ginger started life. We have no idea how many generations of ferals had lived under the school buildings where we first met him and his colony mates, but it's safe to say Big Ginger was well and truly suspicious of people. As were most of the other adult cats. Four of the females gave birth to litters of kittens while in the rescue center—judging by the telltale ginger and tortoiseshell coats among them, we assumed Big Ginger was the father of at least some of them. Despite their antisocial dad, these kittens were young enough to be introduced to people and socialized at the rescue center before finding homes. This would not have been possible for Big Ginger. He would never be able to tolerate living in such close proximity to humans, although as time went on, he gradually accepted my daily presence outdoors at the farm and would sit politely some distance away waiting for his dinner.

The Origins of the Domestic Cat

Where did it all begin? It is only really in the past twenty years that we have discovered the true origins of today's domestic cat. Prior to that, from the many artistic portrayals of cats on ancient Egyptian tombs and temples from around three and a half thousand years ago, we simply knew that a special relationship with cats existed at that time. Images of cats sitting under people's chairs or on their laps led to the assumption that it was the ancient Egyptians who first domesticated the cat. But which "cat" did they domesticate? And did cat domestication only occur in ancient Egypt?

The first step toward finding the answers to these questions came in 2007, when a study of the DNA of the entire cat family—the Felidae—revealed that it was composed of eight distinct groups or lineages. These groups diverged from their common ancestor, the catlike Pseudaelurus, at different times, beginning with the *Panthera* lineage (containing, among others, lions and tigers) over ten million years ago. The very last group to branch off the family tree, around 3.4 million years ago, was a lineage containing various species of small wildcat—the *Felis* lineage. From genetic comparisons within the study, researchers found that the domestic cat fit within this lineage.

It seemed likely, then, that the domestic cat evolved from one or more of these species of wildcats. A groundbreaking study by Carlos Driscoll and colleagues pinpointed the identity of this ancestor. In a huge project comparing genetic material from some 979 domestic cats and wildcats, Driscoll and his colleagues discovered that all of today's domestic cats are descended from the African wildcat (sometimes also referred to as the Near Eastern wildcat), *Felis lybica lybica*. This raises a very big question: Why, out of the forty wild species of cat within the Felidae, did only one become domesticated?

Humans have always been fascinated by felids. All types of them—from large, roaring ones to small wildcats. Long before we had any domestic cats, we were taming wild species of cats all over the world. In a huge review of literature on the subject, Eric Faure and Andrew Kitchener estimated that almost 40 percent of all felid species have at some point been tamed by humankind. In many cases this was to help us with hunting, from the removal of pest species, such as mice, to the capture of prey, such as gazelle, for our own consumption. Others have been tamed for sport—the caracal, for example, was frequently "put among the pigeons" in India with bets taken on how many it could fell with one sweep of its paw.

There seems to be a mixture of tamable and untamable species across the different felid lineages. Tame felid species are dotted around the world but are especially concentrated in civilizations where, historically, cats and other animals possess cultural significance. For example, the jaguarundi is just one of many wild animals tamed since pre-Columbian times by Amazonian societies as rodent-catching "pets"—most of whom were raised by humans as cubs after their mothers had been killed.

Apparently one of the easiest wildcats to tame is the beautiful, elegant cheetah. Some historians suggest that a human-cheetah relationship began with the Sumerians taming them as long as five thousand years ago. The ancient Egyptians certainly kept cheetahs for hunting, and they also believed that the animals helped carry the souls of pharaohs to the afterworld. The cheetah's affiliation with people continued through the centuries, with references to it as an excellent hunting companion to Russian princes in the eleventh and twelfth centuries, and to Armenian royalty in the fifteenth century. The fashion of using cheetahs for hunting gradually spread throughout the European nobility—their "hunting leopards," as they called them, rode seated on horses behind the huntsmen. In India, Akbar the Great, Mogul emperor from 1556 to 1602, was particularly preoccupied with cheetahs, often training them himself.

And yet these elegant, leggy cats with their luxurious spotted coats never became domesticated. A hint as to why was provided by Jehān-gīr, son and heir of Akbar the Great, who, in 1613, wrote in his memoirs: "It is an established fact that cheetahs in unaccustomed places do not pair off with a female, for my revered father once collected together 1,000 cheetahs. He was very desirous that they should pair, but this in no way came off."

Jehān-gīr proved to be right: cheetahs are notoriously hard to breed in captivity. Even zoos struggled to make a success of it until the early 1960s. With cheetahs too shy to breed near people, their domestication was something of a nonstarter. The cheetahs kept by societies throughout the millennia were, instead, all tamed individually.

The ancient Egyptians were particularly skilled at taming different species of cat. Along with cheetahs and African wildcats, there is evidence that they also tamed caracals, servals, and the local jungle cats (*Felis chaus*). Since there is no genetic hint of any of these in our modern-day domestic cats, these associations must have fizzled out fairly quickly. No one can really explain why—were these other felid species, like cheetahs, less amenable to breeding near people than wildcats? Or perhaps they were just not as friendly.

Given that the African wildcat was so successful, people have wondered about its closer relatives and why there is no evidence of their genes in domestic cats today. European wildcats (*Felis silvestris*), for example, are very similar in size and appearance to their African cousin, and doubtless are equally efficient mousers. Why didn't we domesticate them? From the records of various people's attempts to tame them, the answer becomes evident. The northernmost version of the European wildcat lives in Scotland. Here, British wildlife photographer Frances Pitt, writing in 1936, documented one of her attempts at taming Scottish wildcat kittens. "Then Satan arrived. He was but a wee scrap of yellow-grey-tabby fur, as small a kitten as I could desire; but his name was bestowed on sight and never changed." His name says it all—Satan was untamable.

In contrast to Satan the Scottish wildcat and his European relatives, there are other felid species that are now known to be eminently tamable, yet for which historical records of tamings are rare. Despite being good candidates for taming, some species inhabited parts of the world that didn't coincide with the emerging ancient civilizations. The lynx is a classic example of this—these cats were just never in a place where people found them useful, other than as something to hunt for fur and food.

And so, despite having so many other felids to compete with for our affections, African wildcats were the ones that traveled the world, making it over our doorsteps and into our homes. Their prowess at hunting, small size, and ease of transportation (on land or by sea), combined with their tamability, were ideal characteristics. Equally important was their location around growing human communities, where their presence proved useful. They were simply in the right place at the right time—with the right qualifications.

How and Where Did It Happen?

Driscoll and coworkers' discovery that the African wildcat is the lone ancestor of the domestic cat opened the door for other scientists to examine this relationship in even more detail. Subsequent studies, delving into genetic and archaeological evidence, revealed that the gene pool of our modern-day cats actually consists of ancient genetic material handed down from two geographically distinct populations of this wildcat. One of these populations was, as originally suspected, located around Egypt. The other came from farther north, in an area of the Near East known as the Fertile Crescent, also called the "cradle of civilization."

These two inputs into the domestic cat gene pool appear to have occurred at different times—the Fertile Crescent input occurred much earlier than the Egyptian one (probably about three thousand years before). Despite this, once the Egyptian cat started to spread, it became the more abundant subtype of the two.

Suddenly the story was far more complex than simply "The ancient Egyptians domesticated cats." Piecing it all together, we can see how the cat's domestication journey may have happened.

Around ten thousand years ago, groups of Neolithic human hunter-gatherers from the plains of the Fertile Crescent began to experiment with growing crops from grains. As they learned how to harvest and store grains successfully, and the need to wander far to hunt and gather waned, early settlements developed. Farmers began to expand their thinking beyond crops. They capitalized on the presence of local wild animals that could be caught and bred for food, milk, skins, and coats. Ancestors of today's goats, cattle, and sheep were gradually domesticated in this way. These "barnyard" animals all had "domestic" qualifications in common: they already lived in herds and so were social animals able to tolerate being penned in together; they adapted readily to eating the sorts of food available; and they instinctively followed a leader, a role taken over by the farmers who would also have controlled their breeding.

Hovering around the fringes of these early farming villages were some small, opportunistic observers—African wildcats. Naturally solitary, hunting alone in their own well-established territories, they would normally instinctively avoid interaction with one another as much as possible, apart from an occasional liaison for mating purposes. Communication would have been long distance, via scent deposits left as they went about their business. However, hungry and curious, they were attracted to these new human settlements by food, possibly to glean scraps of meat from the discarded piles of bones thrown out by villagers, and also to harvest the increasing numbers of rodents inhabiting the farmers' expanding grain stores. These rich local concentrations of food were big enough for wildcats to share, and so groups of them may have begun to hang around, on the lookout for their next meal. As they lurked around the villages, the wildcats would have inevitably come across one another much more than in their normal ranges outside human habitations. Although scent is a great way to communicate at a distance, up close, signals that were more instant and obvious were required in order to avoid confrontation. They needed to find new ways of communicating among themselves.

From the farmers' perspective, these wildcats were the polar opposite of the barnyard species in terms of suitability for domestication. They had no social skills, they would eat only a very specific meat diet, and they most definitely did not obey orders. Despite there being little chance of ever herding such wildcats, early farmers may have tolerated them as they realized they were catching mice and thereby providing some form of free pest-control service, one their domestic descendants still offer to this day.

So began a tentative mutualistic relationship between humans and wildcats. As with all populations of species, inevitably some of these wildcats would have been braver than others, willing to tolerate closer proximity to both other wildcats and people in order to access this newfound food supply. These tamer wildcats may have naturally bred among themselves, outcompeting wilder individuals for the best food, thereby perpetuating the tendency for tameness in their offspring. This natural selection of friendliness toward humans may have been the beginnings of the process of domestication.

Such wildcat-human relationships may have initially sprung up around various settlements within the Fertile Crescent. Evidence from archaeological and genetic studies suggests that as these Neolithic populations moved on to new areas, the wildcats followed them across the land to reach new settlements in some areas of continental Europe four to six thousand years ago. The point at which people began inviting the tamer wildcats into their homes remains hazy, however, and may not even have happened at this stage of the game.

A similar process occurred in ancient Egypt, where the little local wildcats were tamed, possibly for use as pest controllers, keeping mice, scorpions, and snakes at bay. However, things took a slightly different path here compared with the lowly cats of the Fertile Crescent. As well as acting as pest controllers, Egyptian cats became associated with various ancient Egyptian deities, in particular the goddess Bastet.

Bastet

The importance and the reverence shown toward cats grew, culminating in laws being passed forbidding their harm (killing a cat was punishable by death). They became increasingly valued for their companionship and were kept as pets; the natural death of a pet cat resulted in an elaborate burial and the whole household shaving off their eyebrows as a mark of respect. Beautifully painted depictions of cats in domestic settings show that the wildcats very much had their paws in the door of some households by three and a half thousand years ago.

Over at the temples, however, life for cats was not so rosy. Worshipping of deities, it seems, required substantial and frequent offerings to keep the gods and goddesses happy. For Bastet, offerings took the form of mummified cats. In odd juxtaposition to their widespread protection and adulation, cats were also bred en masse at the temples in large catteries, specifically to be sacrificed at a young age, mummified, and sold as offerings to the goddess. But some of these young cats must have been spared in order to breed subsequent generations. In an enlightening foray into the world of cat-human alliances, Eric Faure and Andrew Kitchener liken this breeding of generations of tame wildcats in rapid succession and in a captive environment to Belyaev's silver fox experiment. In what Faure and Kitchener describe as "an accident of history," the temple Egyptians may have unwittingly rapidly produced a more domesticated version of the wildcat. Whether any of these temple cats ever survived or escaped to mix with the household populations remains unknown, but it seems possible that the temple guardians would have had their favorites and kept them as pets.

The close, crowded confines of the temple cattery must have created a need, even stronger than among the village wildcats of the Fertile Crescent, for effective new methods of communication between cats. These groups may have been the starting point for the development of new cat-to-cat signals—easy-to-see visual cues such as different tail postures, along with more tactile signs such as rubbing and grooming each other. Later chapters explore how these signals have developed, how cats use them to communicate with one another, and how they have adapted them to communicate with people too.

Around the World in Various Ways

While the Near Eastern population of wildcats followed their people on land, those originating from Egypt found an additional, faster way to spread around the Old World—ships. Despite a law against the export of cats, many escaped the confines of Egypt stowed away aboard ships bound for trading along the routes of the Mediterranean. Cats were the perfect hitchhiking guest; they earned their keep catching and eating a new persistent pest, the house mouse. Other than being thrown the odd fish, cats didn't need much feeding or even access to water, as they gained enough moisture from their rodent dinners. Small and unobtrusive, they soon became the sailors' friends. Leaving the protected environs of ancient Egypt, shipboard cats were still generally highly respected. Everywhere they landed, there seemed to be some precious commodity that people needed protecting from rodents—from silk moth cocoons in China to manuscripts in Japan to grain stores in Greece and Italy.

It sounds simple—as if cats just found job opportunities in each country and filled them. But it wasn't all smooth sailing; there was competition on many of the new shores that they reached. These maritime domesticated wildcats from Egypt would have disembarked to find other local species already fulfilling the pest-control role. In China, for example, scientists have found evidence from the Neolithic period of leopard cats (*Prionailurus bengalensis*) living in association with people. No evidence of the leopard cat exists in the genetic makeup of modern-day cats of the region, however, so the African wildcat may have gradually nudged this other cat species from its relationship with humans. The Greeks and Romans were tough to win over, as they already had excellent pest controllers in the form of various mustelid species such as polecats and weasels. Notoriously good at rodent control, these species were nevertheless eventually displaced in their role by cats, despite the latter purportedly being less proficient at the task. The reason remains unclear—possibly the mustelids were more aloof and less receptive to people than cats.

And so, cats spread. They picked up associations with new deities—Artemis in Greece, Diana in Italy, and the Norse goddess Freya. From around 500 BCE to AD 1200, cats began to pop up throughout European countries. They followed the Romans to conquer their empire, then hopped on board with Vikings as they navigated the seas and rampaged their way through new lands. Gene mutations meant new colors and patterns began to appear in cats' coats—orange, black, white, and, later, a new blotched type of tabby, different from that of their striped-tabby ancestor.

It is hard to tell just how much of a pet domestic cats were by the time of the first millennium AD. Their reverence in ancient Egypt must have been hard to replicate in any of their new locations. In Europe, certainly, the utilitarian relationship with humans most likely continued, with cats hanging around as mousers. In this respect they held a certain value, albeit a monetary rather than sentimental one—Hywel the Good, King of South Wales, did much to protect the fortunes of Welsh cats in this respect by passing a law in AD 936, outlining a surprisingly detailed price plan for cats. Newborn kittens would each be worth a penny before they even opened their eyes. Thereafter two pennies until they had managed to kill mice, when they would be worth four pennies. This adult cat price, back then, was equivalent to the price of a sheep or goat, raising the profile of the humble cat considerably.

It was not all fun games of cat and mouse for the cats of these times, however. As well as being hunters, the growing penchant of people to wear fur meant many cats were skinned for their pelts. Evidence shows younger cats were chosen for this, possibly because their pelts would have been softer and still relatively free of damage or disease. Meanwhile, things were changing in Europe. Christianity was spreading, and with it came an increasing intolerance for "pagan" worship. Being associated with goddesses such as Diana was suddenly not such an advantage for cats. Whispers and rumors spread, and cats, especially black ones, became associated with evil spirits and eventually the devil himself. Women became accused of witchcraft, with cats decreed as their evil sidekicks, or "familiars." Amid this rising hysteria, in 1233 Pope Gregory IX announced his Vox in Rama, giving his sanction for the extermination of all cats. From the thirteenth to seventeenth centuries cats were massacred mercilessly across Europe. Women accused of witchcraft were relentlessly persecuted, tortured, and burned at the stake along with their cats, whose equally tortured fates also included being spit-roasted, thrown from high towers, and burned alive in wicker baskets.

In Spain cats were used for their meat, illustrated by one of the earliest Spanish cookbooks, *Libro de Guisados*, written in 1529 by chef Ruperto de Nola. In a lengthy list of dishes, nestled among the intriguing-sounding "Pottage of Sheep's Trotters," "Dish for the Angels," and "Barding for Peacocks or Capons" is recipe 123 for: "Gato Asado como Se Quiere Comer." Loosely translated as "Roast Cat as You Wish to Eat It."

It was during this same period, in the late 1400s and early 1500s, that some more fortunate Spanish cats managed to escape and hitch a ride aboard Columbus's ships to sail to the New World. Later, in the 1620s to 1640s, more ships and cats set sail from England with the Pilgrims to reach the shores of the New World. Despite the fresh start, witch hunts claimed many cats in these new American populations too. Still cats spread, reaching eastern Australia with European settlers in the 1800s.

Back in England, in case being burned at the stake, thrown from towers, and burned in baskets wasn't enough, cats (and dogs) were blamed for spreading the London plague of 1665. Thousands of them were slaughtered. Only later was it realized that rats were the main transporter of the fleas that carried the plague. Perhaps having a few more cats around would have been a good thing.

The cat soldiered on. The Renaissance, while not exactly a period of rebirth for the cat, brought occasional signs of change amid the continuing witch trials. Cruelty here and a little kindness there, never better personified than in the English nursery rhyme of the late sixteenth century:

Ding dong bell, kitty's in the well
Who put her in? Little Johnny Flynn
Who pulled her out? Little Tommy Stout
What a naughty boy was that, try to
drown poor kitty-cat,
Who ne'er did any harm
But killed all the mice in the farmer's barn!

Gradually, attitudes toward cats improved in the late nineteenth and twentieth centuries. Artists began to feature them in paintings, while cat-loving writers such as Christopher Smart and Samuel Johnson extolled their virtues in their books and poems. Cats became fashionable once more, and this renewed interest in them led to the beginning of cat breeding, or "cat fancy"—deliberately selecting pairs of cats to mate in order to produce kittens with particular appearances. Despite their popularity, as mentioned earlier, these pedigreed cats are still very much in the minority, vastly outnumbered by the far more abundant random-bred house cats, or "moggies," that inhabit our homes, streets, towns, farms, and countryside worldwide.

The Sometimes-Social Cat

And so, can we really say, after all the love and hate and torture, that we have domesticated the cat? We now know roughly where and when cat-human associations began, but much debate continues over how or even *if* the wildcat became properly domesticated. Early wildcat-human associations in the Fertile Crescent and ancient Egypt undoubtedly led to stronger associations as time went by. However, in contrast to our deliberate domestication of other species, our relationship with wildcats was more casual—we didn't need to control them to capitalize on their prey-catching skills. It is likely that wildcats drifted toward domestication simply through their own behavior. As the tamer individuals ventured into human villages and bred among themselves, their kittens would have benefited from the food and shelter offered by human habitation.

In this way, the friendliest wildcats gradually and quite naturally began to domesticate themselves, with minimal input from humans. This has been described as "self-domestication," following a model proposed for our other best friend, the dog, and, intriguingly, for us humans too. Just as friendlier wildcats and wolves outcompeted less friendly ones, Brian Hare from Duke University suggests that *Homo sapiens* outcompeted other hominids of their time by learning to be more approachable—a true "survival of the friendliest."

Cooperation is the name of the game. For humans it was a no-brainer—they simply had to learn to work well with other humans. The "protodog" wolves were already used to socializing with their own kind prior to domestication. They then creatively learned to adapt these skills for living and working with humans. But the solitary wildcat had to learn to communicate not only with a new species—humans—but with each other too. A double challenge. So, while we were learning to chat with one another and beginning to teach our dogs new tricks and skills, the lifestyle of the wildcat underwent a sea change from loner to socializer. They worked out how to use signals that other cats could see or feel up close. They also discovered they could tap into the human love of vocal conversation, developing their existing vocalizations into ones that, just like dogs with their barks, gained our attention. As Driscoll and coauthors observed, "Cats are the only domesticate that is social under domestication yet solitary in the wild."

Domestic cats, wise as ever, have kept their options open. Rather than completely changing to become a social species, they have retained the ability to live a solitary or social life according to their circumstances. As such they are often referred to as "facultatively social" or "social generalists." At one end of the spectrum, our well-fed house cats may live as single cats or in pairs or small groups in our homes. These are the cats that most have to develop their "people" communication skills, as well as potentially dealing with feline housemates and neighborhood cats, too, if they are allowed outdoors. Sadly, many cats, for varying reasons, lose the security and comfort of their home and become strays. They may adapt to life on the streets with or without other cats for company or sometimes, if they are lucky, find themselves a new home. Others, just like Ginger and his friends, after several generations of life separated from people, are far from tame, but where food supplies allow, they may form larger free-living groups, or colonies. Studies of these have shown that a social system exists based around related mothers reproducing and pooling their litters, with males living more independently on the periphery. These are the situations in which cat-to-cat social skills are so important. Even in neutered colonies, like mine, without the ties of breeding, interactions are far from random, with many cats having preferred partners with whom they associate.

This ability to chop and change from solitary to social life has been a key part of the domestic cat's success. Behind it lies their enviable talent for developing new methods of communication. The following chapters delve into the scientific discoveries of how cats have learned to step out from their world of smells, using visual, tactile, and sound signals to get their messages across to both each other and humans.

Are cats as domesticated as they will ever be? Or are they still on their domestication journey? It is a conundrum that is almost impossible to answer. Cats display many domesticated features in common with dogs and other domesticates, such as increased tolerance and sociality, but dogs are clearly further down the domestication route than cats. The canine eagerness to please, for example, is a trait rarely displayed by cats. Maybe cats, too, will one day hang on our every word. Let's not hold our breaths.

Arriving one day at the farm, I realized that a new normal had emerged for the cat colony. They had become used to me being around and didn't scatter far and wide on my arrival. Instead, they carried on with their business of interacting with or avoiding one another as they wandered in and out of the adjacent woods. Even Big Ginger seemed tolerant of my presence, provided I kept my distance. I posted myself in my usual place up a little hill, comfortably far enough away from the group, and settled down with my tape recorder, binoculars (I didn't want to miss a thing), and notebook. Before I left the farm that day, I walked down and gently lifted the lid on the apparently deserted catterama. Peeking in, I smiled to myself as I spotted telltale ginger hairs on one of the old blankets I'd left inside. Feral and unsocialized he may have been, but Big Ginger clearly was not averse to a few domestic home comforts.

CHAPTER 2

MAKING SCENTS

A man has to work so hard so that something of his personality stays alive. A tomcat has it so easy, he has only to spray and his presence is there for years on rainy days.

—Albert Einstein

Sheba, the rescue cat I was fostering, finished washing her sixth and final kitten born that morning. As she lay back exhausted, I watched anxiously as her six tiny and still blind little babies began to snuffle and root around hungrily in her fur. How could they possibly find their way to food? I needn't have worried; within five minutes they were all lined up at the milk bar, suckling and snoozing contentedly. For me, the miracle of birth was almost trumped by this second miracle of the morning—how did they manage that all by themselves?

While kittens' eyes are still shut and their hearing is in a rudimentary stage of development at birth, their senses of touch and smell (olfaction) are already up and running. These are the senses they use as they begin to nuzzle their way across their mother's soft underbelly (or ventrum) and attach to the first nipple they find. Within just a few hours, they become a little more discerning and show a preference for the teats toward the rear of their mother's body. It remains a mystery why they prefer these, as research has confirmed that the rear teats provide the same level of nutritional sustenance as those positioned near the front. Nevertheless, a competition for the back rows develops involving a considerable amount of pushing and shoving among the kittens. Within a few days they come to some sort of teat order arrangement, or teat "ownership," each kitten using mainly one or sometimes two particular nipples from which they subsequently nurse. Despite not being able see at this stage, the kittens are guided by scent to return to their own special nipples regardless of how the mother cat positions herself or whether their littermates are there for them to climb over. This pattern of suckling behavior is very different from that of dogs. Puppies show no tendency to stick to one nipple when nursing, often swapping between nipples during a nursing session.

A kitten will nurse from a female cat that isn't their own mother as long as she is producing milk, a fact that is put to good use when female cats pool their litters in farm cat colonies and other group-rearing situations. However, when researchers took kittens that had developed a preference for a particular teat on their own mother and put them to nurse from a different lactating female, the kittens did not instinctively find and use a teat in the same position. They seem to recognize specific odor signals that lead them to the appropriate teat on their mother rather than learning its position. A maternal pheromone may enable the initial navigation to the teats, followed by a gradual learning process whereby the kitten becomes familiar with the distinctive smell of their favorite teat. There are many sources of smell in any nest of kittens—not just the mother's milk but also saliva or skin gland secretions from both the mother and the kittens. As the kittens root through their mother's fur and latch onto and off their chosen nipple, they may create what is to them a very distinctive individual milky, salivary trail that leads them back each time to the same teat.

During the early days in the nest with their mother and siblings, kittens learn their mother's scent. They appear to remember it into adulthood too. A paper whose title starts intriguingly with "Are You My Mummy?" describes the fascinating results of an odor memory study in kittens. Researchers presented eight-week-old kittens, who had recently been separated from their mother after weaning, with three cotton swabs. One smelled of their mother, one of a different unknown female cat, and the third was blank. Unexpectedly, the young kittens, rather than sniffing most at the scent from their mother, showed more interest in the smell of the unknown female. Possibly there was some novelty effect going on here, with the smell of an unfamiliar, new cat being more interesting to the kittens whose olfactory experience to date had been restricted to littermates and their mother. Fortunately, the researchers didn't stop there. They kept track of the kittens when they went to their new homes and visited them at four, six, and twelve months old. As before, they presented them with three different scents, one from their mother, one from a strange female, and one blank. At four months old the kittens showed mixed responses to the smells, with no overall preference in evidence. However, by the time they were six months old and again at twelve months, the kittens spent a significantly longer time sniffing the scent from their own mother compared with the other two swabs. Did they know it was their mother? Possibly, although the authors point out that these older kittens may have simply recognized the smell as being a familiar one, and therefore spent longer sniffing it. Perhaps this is similar to when we catch a scent on the air and stop to wonder what it reminds us of. Whether the kittens recognize the scent as their own mother's is difficult to confirm. However, the study does show that a mother cat has her own individual odor that must be consistent enough despite changes in her reproductive state over time for her kittens to remember her scent ten months after they last saw or smelled her.

Mother cats can tell which kittens belong to them, too—when litters are mixed up they use their kittens' scent to distinguish them from offspring of other mothers. Despite this, when faced with a selection of kittens who have strayed from the nest, her own and others that aren't hers, a mother cat doesn't appear to favor her own offspring when retrieving them. The reason for this is uncertain, although distress vocalizations from kittens that are lost from their nest are known to be very powerful, so it may just be hard for the mother to resist retrieving them, regardless of whether they are hers. In the wild, a squeaking kitten out in the open is likely to attract predators, which is bad news for any other kittens around it. A rapid rescue of any crying kitten would be a good strategy to prevent them from drawing unwanted attention.

An investigation by Elisa Jacinto and colleagues found that kittens within a litter develop individual odor "signatures." They looked at adult cats' sensitivity to different kittens' odors using what is known as a habituation/dishabituation test. This consisted of first presenting a cat with the odor of a kitten (collected by rubbing a cotton swab over the kitten's body) and then recording how long the cat sniffed the scent. This was done twice more with odor from the same kitten. Known as the habituation phase, sniffing time tends to decrease with subsequent presentations of the same smell. After this, the adult cat was presented with a cotton swab bearing the scent of a different kitten (the dishabituation odor). If the cat detected a difference in odor between the first and second kitten smells, the rate of sniffing of this new smell should increase in relation to the last presentation of the habituation scent.

The results showed that adult cats of both sexes were able to tell the difference between the scents of individual kittens from the same litter, although not until the kittens had reached seven weeks old. However, surprisingly, this didn't apply to mothers sniffing the scent of their own kittens. The researchers suggested that a mother cat may learn and respond to an overall odor of her litter or nest that is carried on each kitten, rather than to their individual odors. Alternatively, she may know all their individual scent profiles but respond to them as a group as simply "belonging to her." Possibly this simplifies things for a mother cat keeping track of her kittens on a daily basis.

Scent remains a hugely important part of life into adulthood for cats. They investigate most things that they come across, whether it's food, another cat, a person, or any object, first by smelling them. When two cats meet amicably, they sniff each other either face-to-face (with one of my favorite cat behaviors, "touch noses"), or they may investigate each other's body or rear end instead, gaining olfactory information about the other cat. One study of free-ranging cats observed twenty-two different social behaviors, with sniffs comprising 30 percent of the total behaviors recorded. In preference to the auditory or visual methods favored by humans, cats, like many carnivores, rely heavily on scent signals for

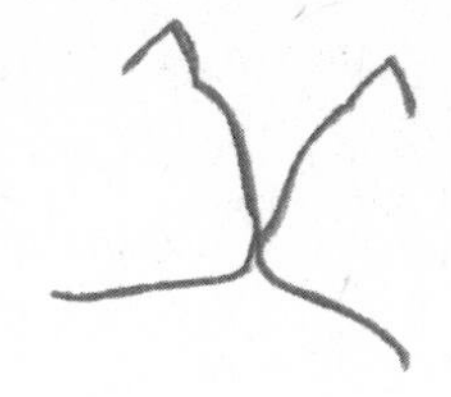

communication, using them to mark their territories and advertise their sexual status to potential mates. Scent marks convey relatively long-lasting information about the scent depositor without the need for face-to-face encounters, a legacy from their solitary living ancestor, the wildcat. As well as depositing secretions from specific glands on their bodies, some cats use their urine and feces as signals for other cats to interpret, often choosing sites for their deposition that will maximize the signal's durability and potential for discovery.

Does It Taste as Good as It Smells? The Joy of an Extra Nose

The domestic cat has an impressive sense of smell, partly due to the size of the surface area inside the nose devoted to receiving odor particles. This olfactory surface (or membrane), while only around two to four square centimeters in humans, is a whopping twenty square centimeters in cats, wrapped over the convoluted folds of a bony maze inside their nose—far more space for smells to make themselves known. Information from the smell receptors in the olfactory membrane is transmitted to and processed by an area at the front of their brain known as the olfactory lobe, or bulb.

Alongside their impressive sense of smell via the nose, cats have an ace up their sleeve—a second method of scent detection that they start to use at around six weeks of age. In the roof of their mouth, a tiny slit behind the upper incisors marks the openings of two narrow tubes known as the nasopalatine canals. These run to a pair of fluid-filled sacs that together form what is known as the vomeronasal organ (VNO), or Jacobson's organ. Packed with chemical receptors, the VNO connects to the brain in a special area separate from the olfactory bulb known as the accessory olfactory bulb. This whole separate smelling setup is known as the vomeronasal apparatus.

The opening to the VNO is in the roof of the mouth, so to enable scents to enter it, cats raise their top lip and open their mouths slightly. This produces a characteristic kind of grimace or gape behavior frequently described by the German word *flehmen,* which means "to bare the upper teeth." Cats will often sniff an item through their nose first, followed by the flehmen response. They take on a kind of dreamy, faraway look as they do this. Unlike in the nose, where sniffed scent molecules simply land on the olfactory surface and trigger receptors, the route to the VNO is more intricate. Molecules need to be fluid-borne in order to navigate the narrow nasopalatine canals leading from the mouth to the organ. The cat appears to make actual physical contact with the source of the scent with its mouth—in effect, it "tastes" the smell as it is transferred via its saliva into the nasopalatine canals en route to the VNO.

For such a small structure, the VNO has caused a considerable amount of debate and controversy since it was discovered. The organ is not unique to cats, having been found in dogs, horses, snakes, mice, and numerous other animals. It was named after Ludvig Levin Jacobson, who, in his 1813 work titled "Anatomical Description of a New Organ in the Nose of Domesticated Animals," presented a description of the organ's structure in a variety of nonhuman mammals, illustrated by drawings from a horse. With such a widespread existence across species, the question inevitably arose as to whether humans have a VNO. Opinions vary widely; however, the most likely answer seems to be that a vomeronasal-like structure begins to develop in human embryos but in adults remains vestigial. It lacks any neural connections to the brain, and there is no accessory olfactory bulb. So, sadly, we lack the ability to "taste" smells in the way that cats do.

Cats investigate most smells the conventional way, by using their nose. Certain smells, however, particularly those deposited by another cat or other animal, will also stimulate them to use their VNO. Sources of such scents include urine and feces, along with surfaces that have been rubbed on or scratched. These scents, described in more detail in the following sections, provide a surprising amount of important social information for cats. Traditionally referred to simply as pheromones, they are now also described as chemosignals, sociochemicals, or semiochemicals. Having two methods of smelling must have been hugely advantageous to the ancestral wildcats that relied heavily on scent not just to locate prey but also to keep track of one another and avoid unnecessary encounters.

The Smelliest Smells

I sat on the edge of the sofa with my notebook resting on my knees and listened as Mrs. Jones spoke. "He does it to spite us, you know," she complained. "I even saw him doing it on my lovely new boots just inside the front door yesterday." I glanced at the cat that was sitting on the back of the sofa, watching intently out the window. Cecil was guilty of the crime that I spent so many of my visits as a cat behavior counselor discussing—urine spraying. I asked Mrs. Jones why she thought he was being spiteful. "Well, because we went away for a week and he's punishing us." Digging a little deeper, I discovered that while his owners had been away, Cecil had been cared for by a kindly neighbor who had popped in once a day to feed him. Mrs. Jones continued, "She [the neighbor] said the food bowls were always scattered all over the place and everything was a mess each time she came. So Cecil was obviously cross with us even then." It occurred to me that possibly someone other than just the neighbor had been "popping in" while the Joneses were away. "Do you ever see other cats near the house?" I asked. "Well, there is that black-and-white stray with the big head that keeps hanging around the cat flap," Mrs. Jones replied. There was our explanation. In the absence of Cecil's human shield of protection, a local opportunist tomcat (an unneutered male) had invaded his small but precious patch of space. Poor Cecil, threatened by this intrusion, had felt the need to mark the inside of the house as his own to try and deter the itinerant tom.

One of the most effective methods used by cats to spread their social scent messages is urine spraying (earning it the delightful nickname of "pee-mail"). This is just fine, if a little smelly, outdoors, where the scent catches the wind for all to enjoy. However, when it occurs in the home, spraying can rapidly become a huge bone of contention between cats and their owners. The act of spraying is very distinctive—the cat stands and reverses up to a vertical surface, raises their tail, and holds it there, upright and quivering, while a spray of urine is directed horizontally onto the target. The spray results in a maximum dispersal of urine onto a surface, at a height most likely to be encountered by other feline noses, followed by further spreading of the liquid deposit as it slides and drips down. Spraying is very different from urinating in a squat position, which cats adopt for simple toileting purposes. Squat urinations are usually covered by the depositing cat scraping soil or litter over it, while spraying is a deliberate marking behavior designed to leave a scent signal for another animal to smell. It is a behavior most commonly associated with unneutered tomcats, who tend to be more preoccupied with advertising their presence, although neutered males and female cats may sometimes spray as well.

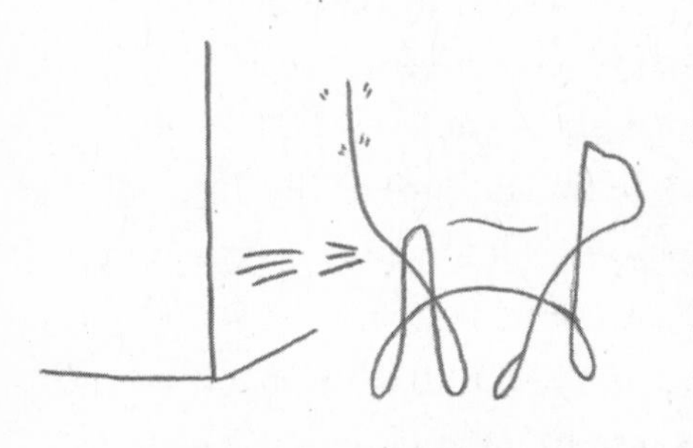

Domestic cats with outdoor access often spray on vertical surfaces most likely to be obvious to other cats, such as fence posts, trees, and the edges of buildings. For group-living cats, urine spraying continues to be an important method of communication, even though closer-range visual methods are available too. In colonies of unneutered cats, such as on farms, the amount of spraying tends to increase in both male and female cats when the females are in estrus (the period when they are sexually receptive). Spraying is not just reserved for sexual and courtship situations, however, as intact males will often spray as they go about their day-to-day activities, patrolling their usual ranges and hunting.

Unfortunately for owners, some pet cats, particularly males, feel compelled to spray around the home, marking their territory just as outdoor cats do. This tends to happen in more competitive or stressful situations, such as when multiple cats live together but aren't particularly compatible, or when other cats from outside unexpectedly enter the home, just as with Cecil. Household renovations, redecoration, and new furniture can all provoke an outbreak of anxious spraying in a resident cat. Cats really don't like change, especially in smells, and urine spraying is usually a good sign that something is worrying them. Just like with a feral cat, an indoor cat will choose conspicuous vertical sites, and they will spray repeatedly. Cupboard doors, door frames, potted plants, and soft furnishings such as curtains are all perfect surfaces. Sprayers will also often target electrical appliances that

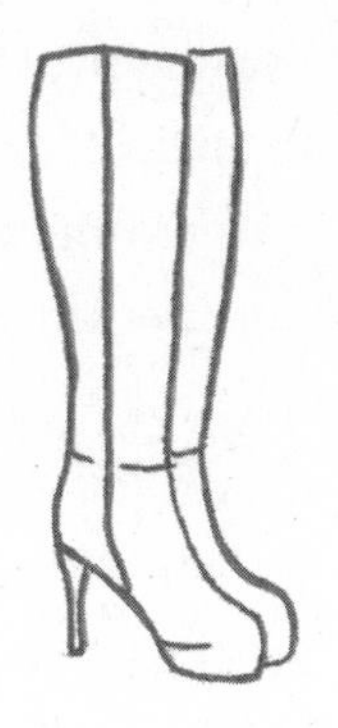

warm up when in operation, such as computers, dishwashers, and, a common favorite, toasters. The heat helps amplify and disperse the odor of the sprayed urine even further, much to the horror of the unsuspecting owner as they switch on their computer for the day or pop some bread into the toaster for breakfast. Items bearing new scents that the owners have unwittingly brought into the house may also be targeted, which is why Mrs. Jones's lovely new boots upset Cecil so much.

What exactly are the messages being conveyed in these liberal splatterings of urine around a house, garden, or farmyard? Observation of the reaction of a cat when sniffing the urine deposited by another suggests that the urine itself is not intimidating to the recipient. Interesting yes, but not scary—the mark just seems to serve as an information point.

Early studies by Warner Passanisi and David Macdonald looked at responses of farm cats to different types and sources of other cats' urine, comparing how much each was investigated—that is, how interesting each urine sample was relative to others. They compared responses to samples of sprayed urine from a male cat, squat urine from a male cat, and squat urine from a female cat. Test cats were offered samples of each type of urine from cats in their own colony, from a colony living directly adjacent to them, or from a completely strange, unknown colony. The results showed that, while cats would sniff squat urination, by far the most "interesting" type in terms of time spent sniffing was the sprayed urine—this was the case for both male and female sniffers. In addition, male cats (and, to a lesser extent, female cats) spent different amounts of time sniffing urine samples taken from the three different source colonies. Those taken from the unknown colony were sniffed the longest, while those sourced from neighboring groups were sniffed for less time. Samples taken from cats in their own colony received the least amount of sniffing. While measurement of sniffing time is a fairly limited method of assessing the information cats gain from urine, the results from this study do suggest that urine from some cats certainly merits greater investigation than samples taken from others.

The different reactions of cats to squat and sprayed urine suggests that there may be something distinctive about the sprayed version. There do indeed seem to be slight differences in the chemical compositions of squat versus sprayed cat urine, although the origin of these is still uncertain. It has been suggested in the past that sprayed urine contains secretions from the cat's anal sacs that are released into the urine as it is sprayed. Research on large felids such as lions, however, has found no evidence of anal sac secretions in their urine marks. Also, given that the anal glands are not in direct contact with urine as it is expelled, it seems unlikely that much could pass from them into the urine stream.

Scientists analyzing the contents of domestic cat urine found its contents are a mixture of substances, both volatile (ones that evaporate into a form of gas and hang about in the air) and nonvolatile (ones that don't). A detailed investigation of the many volatile components of male cat urine revealed that they change constantly as the urine ages, particularly in the first thirty minutes after it has been exposed to the air. The researchers explored whether cats could detect this temporal change by using habituation/dishabituation sniffing tests. They presented the cats with fresh urine from another individual four times in a row, followed by one presentation of older (twenty-four hours) urine from the same cat for the final dishabituation test. They found that, having decreased their sniffing rate over the first four presentations, the cats then increased their rate of sniffing when presented with this final, older sample, indicating that they distinguished between older and fresher urine. What is difficult to establish is how much information the cats can glean when sniffing—do they actually know how old the respective marks are simply from smelling them, or just that they are different?

The same study also confirmed that cats are able to distinguish between the urine of two different individuals, although that becomes more difficult as the urine ages. Quite which components of the urine are used to tell the difference remains something of a mystery.

These scent-perception skills were tested indoors under controlled conditions. Outdoors, in a natural setting, there will be the added challenges of environmental factors such as temperature, rain, and wind, along with the type of surface bearing the mark, all of which may affect the rate at which the properties of urine change. Nevertheless, as we watch a neighborhood cat sniffing a fence post and dreamily performing the flehmen response, we can surmise that they may be deducing whether another cat has passed by and when. In this way, in areas where the cat population is high, cats wandering outdoors can avoid bumping into one another and may even learn to "time-share" routes, using the same spaces at different times of day.

Among the many substances found in cat urine, scientists discovered something surprising: protein. Mammal species on the whole tend not to excrete protein in their urine, its presence often being a sign of illness or disease. Cat urine, though, routinely contains a protein called cauxin. This protein is responsible for regulating the production of an amino acid called felinine, unique to certain members of the Felidae family, which is also excreted in their urine. Felinine is synthesized from cysteine and methionine, two amino acids that cats can obtain only from their meat-based diet.

Kittens begin to excrete felinine at around three months old, the concentration increasing with age to adulthood, when it is higher in intact males (tomcats) compared with females or neutered males. Odorless in its intact form, once felinine is voided in the urine, exposure to air and microbes causes it to begin to break down, releasing volatile sulfur-containing molecules known as thiols. Along with ammonia, these thiols are responsible for the distinctive smell of cat urine, which becomes stronger as time passes and the scent matures.

It has been suggested that the smelliness of a tomcat's urine may serve an important function in the choice of mates by female cats, at least among feral cats that must fend for themselves and hunt to survive. Logically, due to the need to consume meat to produce felinine, tomcats that hunt well will consequently produce more felinine and have smellier urine. Such genuine signals of fitness are common in many animal species, enabling females to choose the best potential fathers for their offspring.

For most humans it's hard to miss that catty smell as you walk into a room—highly distinctive and often overwhelmingly strong, it is the dread of many a cat owner. While there is little doubt that we can smell it, our human reaction to cat urine "messages" is usually less than ideal, as far as the cat is concerned, at least. We don't really get the message, other than hopefully realizing something is bothering our cat. Rather than smelling it with interest as another cat would, our natural instinct is to get rid of it as fast as we can, by washing the area or covering it up with a different aroma. Years of cleaning our houses and effective television marketing have encouraged us to use an array of commercial products to achieve the ultimate clean surface. These therefore tend to be our go-to products when cleaning areas where cats have urinated. Herein lies a problem: many cleaning products contain ammonia, and so does cat urine. So when we cover a urine mark with our "pine-scented" ammonia-based product, an area that now smells to us like a Norwegian pine forest smells to the cat like ammonia from cat urine. But not their own. This unfortunately results in them having an overwhelming urge to mark over the offending smell with their own urine once more. In this way owners often unwittingly enter into a scent "conversation" with their cat rather than managing to properly eliminate the original smelly mark.

A better way to remove the urine mark is to use an enzyme-based product, either a commercial one made specifically for cat urine, or a homemade solution of biological washing powder and warm water. The enzymes in these products more effectively break down the offending components of the urine. To avoid the cat starting all over again with a fresh mark, redefining the area as a sleeping or eating place for the cat can help deter them from repeating the cycle.

Intriguingly, people often comment on the similarity between the smell of cat urine and the aroma of certain plants and foods. This is not simply overimaginative scent perception. It turns out that an odor-producing thiol, similar to the ones in cat urine, occurs naturally in sauvignon blanc grapes and blackcurrant plants. Beer made from certain hops, as well as freshly squeezed grapefruit, also contains this same "catty" element. At high concentrations, this particular thiol molecule produces the overpowering smell of cat urine, but much weaker concentrations give off the fruitier notes found in a nice crisp sauvignon blanc.

While it is probably the last thing about our cats that we would ever be inclined to sniff, feces are extremely interesting to cats. Most cats will defecate in loose ground (or a litter box) where possible and then scrape the substrate over the deposit to cover it up. Unneutered tomcats, however,

particularly those associated with colonies and living in large rural territories, may leave some of their feces exposed near the edge of their territories, presumably as an olfactory signal to other cats. Just as with urine, cats can tell a lot about one another from their poop. In one study, when presented with three samples of feces (their own, some from a familiar individual, and some from an unfamiliar individual), cats spent significantly longer sniffing that of the unfamiliar cat compared with the other two. The strange cat's feces were evidently way more interesting. This fascination waned over time, however; the more the unfamiliar feces were presented, the less interesting they became.

By examining cat feces in more detail (yes, some scientists actually do), another study revealed that, surprisingly, the cat-specific amino acid felinine makes another appearance. Previously thought to exist only in urine, it is now known to enter the feces via bile from the liver. In contrast to its expression in urine, felinine is present in equal amounts in the feces of male and female cats. However, compared with females, male cats also have a significantly higher concentration of the compound 3-mercapto-3-methyl-1-butanol (MMB) in their feces. This, interestingly, is derived from felinine. It appears that some felinine breaks down in the large intestine to become MMB and is excreted in the feces along with the normal felinine. Males appear to break down more felinine to MMB than females, producing the high fecal MMB in males, but the actual fecal felinine levels remain the same for both sexes. MMB levels in feces also change over time. Thus, for cats investigating feces, MMB may provide a way of identifying both how long ago it was deposited and the sex of the depositor. Feces also contain mixtures of fatty acids which may be more individual cat–specific, enabling the sniffer to identify the cat from which they came.

Scratching the Surface

If there's one thing that owners dislike their cats doing almost as much as spraying urine, it's scratching. One survey found that 52 percent of people's cats scratched what their owners deemed to be "inappropriate" items inside the home. Scratching is, however, a totally normal and necessary activity for cats, fulfilling a practical function and providing a subtle means of communication.

Cats' claws are constantly growing, and as layers are replaced from beneath, the dead outer layers need to be loosened and shed. Cats scratch, or "strop," their claws on surfaces to remove these outer pieces of claw known as the sheathes or husks. There is far more to scratching than a simple daily pedicure, however. When cats scratch, they are also communicating in two different ways. The scratched grooves etched into a surface form a visual signal—one noticed by cats and, in the human home, most definitely noticed by people. The other message is more subtle. Between their toes, cats have interdigital glands that, when the cat scratches, leave a scent deposit on the scratched surface. Scientists have analyzed this deposit and identified within it a type of pheromone, comprising a mix of fatty acids and described as the feline interdigital semiochemical.

Feral and pet cats with outdoor access tend to find prominent places to scratch-mark, usually along routes that they take regularly within their territory rather than on the boundaries. Wooden fence posts and other easily scratched vertical surfaces such as tree trunks are often used, with softer-barked trees generally more popular than those with harder barks. Indoors, pet cats also seek out particular qualities when selecting surfaces to scratch, often preferring fabrics, which are easy to grip with their claws. They tend to scratch as they wake from a sleep and like to stretch out as they do so. For this reason, they prefer a sturdy, vertical object that is unlikely to tip. A common target is often, as many cat owners well know, the living room couch or an upholstered chair. Some prefer to scratch horizontal surfaces, choosing carpets or sisal doormats.

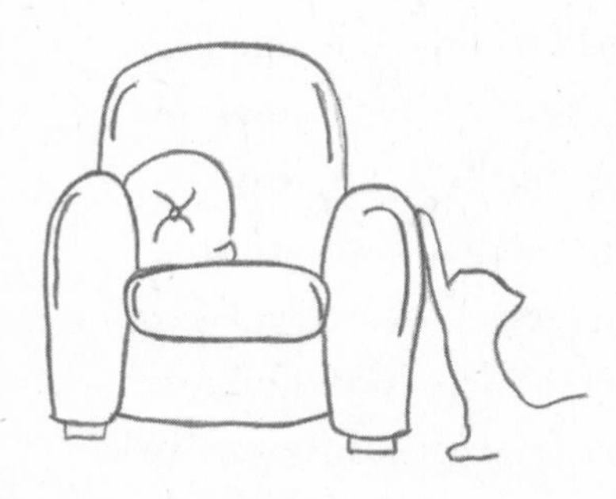

Tension and conflict within the home may bring on more scratching than normal as the cat feels the urge to increase their marking behavior. The common human reaction to yell at the cat to stop scratching therefore rarely has much success long term and can actually make it worse by causing increased stress. Cats may also scratch in front of their owners as a form of attention-seeking behavior, perhaps when hungry or frustrated. Some owners find scratching so unacceptable they resort to declawing their cat (an operation known as onychectomy), often not realizing the scale, invasiveness, and pain of this operation, which in reality involves complete removal of the end joints of the cat's claws, not just the claws themselves. This procedure is now illegal in many countries.

Rather than trying to stop pet cats from scratching, provision of alternative scratching targets is the best solution to the problem. The owner should redirect their cat's scratching to something that they don't mind being scratched and that for the cat is *even better* than whatever they were scratching on before. Ideally, a few good solid scratching posts covered in burlap, sisal rope, or carpet and positioned near where they have previously scratched; next to a place they sleep regularly; or near exits and doorways, where a cat may feel the need to mark its presence. Once a post has begun to be scratched, cats are more likely to return to it because the surface is now rougher and more appealing, and it will have their scent on it. The challenge with many new or replacement scratching items is how to attract the cat to them initially. One way is to spray the post or other item with a catnip spray—for cats that react to catnip, this can be an effective lure, as they may rub their faces on the post or wrap their claws around it in excitement, thereby beginning the process of scratching the new surface.

As discussed more in chapter five, rubbing of cheeks, flanks, and tails by domestic cats both on other cats and on humans appears to fulfill some kind of tactile social-bonding function, similar to mutual grooming. Cats also rub a lot on objects, which may be more of a visual display in certain social settings. However, looking closely at the corner of a cupboard or door frame where a pet cat rubs regularly, you'll often see a smudge. This deposit is the waxy product of the facial glands found on cats' cheeks, temples, and ears, and at the corners of their mouths. A rubbing cat may luxuriously press the whole side of their face from the edge of the chin to the base of the ear along a prominent vertical surface, sometimes back and forth several times and occasionally extending the rub along the length of their body. One more gland, the perioral, situated under the chin, is useful for rubbing on objects that are lower to the ground.

A cat encountering the rub mark of another will investigate it, sometimes at length, sniffing and possibly performing a flehmen response. Female cats in estrus rub more frequently than normal, and male cats show a corresponding greater interest in the rubs of such females. This reaction suggests that rub marks convey information on the sexual status of female cats, doubtless alongside other social details still to be elucidated by science. Presumably, rubbing by cats on a person's leg combines an affectionate and social-bonding gesture with the opportunity, as

Mark Twain so beautifully put it, to "write her autograph all over your leg if you let her."

Following Our Noses

As I unlocked the back door, stepping out to enjoy the warm July afternoon sunshine, Bootsy, who had been watching me in the kitchen, took the opportunity (as she always did) to slip out through the door rather than use the cat flap. Standing on the deck outside, I paused and sniffed—a neighbor was mowing their lawn and the smell was intoxicating. The grassy aroma evoked a brief happy childhood memory of watching my dad mow the lawn, making his customary neat, straight green stripes. I watched Bootsy wander off down the garden sniffing the air, the ground, and the various plants along the way—what smell extravaganza was she experiencing?

In the mammalian sensory lineup, the sense of smell is unique because the information received by the olfactory bulb is transmitted directly to the limbic system of the brain, including the amygdala (which is strongly associated with emotions) and the hippocampus (which controls memories). We all have those freshly mown grass or warm baked brownies moments that send our minds catapulting back to some long-forgotten memory of our past. Scientists call these odor memories, or Proustian moments after the French author Marcel Proust, who first penned a description of such an experience. While it is difficult to know whether cats enjoy these same sensations, they do seem to have the ability to recognize scents from their past, such as that of their mother, as described earlier.

There is a long-held view that humans have a terribly poor sense of smell. This is partly because we compare ourselves with mammal species who rely more on their olfactory abilities on a day-to-day basis. We use "sniffer" dogs, with their supreme scent perception and tracking talents, to root out things we can't smell ourselves. Watching our cats, like Bootsy wandering down the garden, we see that they, too, can apparently smell a whole plethora of scents that pass us by, including those that they leave on us when they rub affectionately around our ankles.

In 1879, the neuroanatomist Paul Broca, famous for his work on the brain and speech, observed that the olfactory lobe of humans was small in relation to the rest of the brain, when compared with that of other animals. Broca claimed that the lobe had atrophied to allow for our much larger frontal lobe. Humans became classified as "microsmatics," or "tiny smellers," with a poor sense of smell, and the label stuck.

Modern-day scientists such as John McGann have begun to challenge this concept, describing it as a "19th-century myth." New isotropic fractionation techniques have shown that, although relatively small, human olfactory lobes contain roughly the same number of neurons as the lobes of "supersmellers." Also, the olfactory part of our brain did not actually shrink over time; the rest of the brain simply expanded, so it became proportionally smaller.

Humans, it turns out, smell better than we tend to think, albeit nowhere near as impressively as cats or dogs. Scientists have discovered that people can discriminate over a trillion different olfactory stimuli and have smelling skills to rival that of other, more scent-focused mammals. One of the more entertaining examples of this was provided by an experiment designed by a team of neuroscientists and engineers from the USA and Israel. Student volunteers had their eyes and ears covered up, and then were put in a field to see if they could follow a scent trail that the experimenters had laid. The scent was chocolate-flavored essential oil, dribbled enticingly in a roughly thirty-three-foot-long trail. Down on their hands and knees, two-thirds of the volunteers successfully completed the task, zigzagging their way along the trail just like dogs do when tracking a scent. And this was with no practice or training at all. Although we will never match the olfactory skills of our dogs or cats, it's not a bad start.

Of great interest to scientists is whether humans, like so many other mammals, use smell in social situations to communicate with one another—and whether olfactory communication occurs between species too. In the Western world, social smells and sniffing have become somewhat taboo among humans. To deliberately sniff someone you meet is deemed inappropriate, something that "animals" do. And yet research suggests that smell is far more important socially than we realize. For a start, just like cats, our bodies are covered in glands that excrete sweat, sebum, and scent. Humans, as cats will testify, are really quite smelly. And indeed, we spend an inordinate amount of time and money masking our own bodily scents with soaps, deodorants, and perfumes.

Research into humans' use of smell is on something of a voyage of rediscovery. For example, we now know that experiencing different emotions causes corresponding changes in our scent—people can tell from someone's odor whether they have just had a fearful experience. We don't tend to see people sniffing themselves or each other, but studies have revealed that this might not be because we don't do it—it's just very subtle. A survey of four hundred people from nineteen different countries asked whether they ever smelled their own hands or armpits. Over 90 percent admitted to each. Ninety-four percent also smelled their close relations, and 60 percent said they smelled strangers.

Alongside all this covert but conscious sniffing of ourselves, it appears that we sometimes also sniff ourselves and others without realizing it. One fascinating experiment focused on the human interactive behavior of shaking hands and whether this was a way humans secretly trade chemosignals. The experimenters found that, following a handshake with another person of the same gender, people sniffed their own right-shaking hand twice as often as normal. They also discovered that during handshakes, several types of volatile molecules, potentially chemosignals, transfer from one person to another.

We obviously have plenty of information to share, whether consciously or not, on our hands. It's important that we bear this in mind during interactions with cats. One of the best ways to initiate a conversation with a cat is to gently hold out a hand toward them, preferably with the fingers curled under slightly so they don't feel like you are reaching to touch them. Give them the opportunity to approach your hand and sniff it—and let them take as long as they need to do this. They will often have a good long sniff, sampling all the smells you have to offer.

Studies on both dogs and horses have shown that these domesticated species are able to distinguish between human body odor sampled after happy versus fearful situations. While this has not yet been tested with cats, it seems likely that they, too, can pick up on the subtleties of our mood via olfactory cues.

Can we smell our cats? One small-scale study of this subject suggests not. When presented with the scent of their own cat versus that of an unfamiliar one, the success rate of the owners was no higher than what would be expected by chance. This contrasts with a similar study that explored whether owners could identify their dogs by smell, which showed that 88.5 percent could. Possibly cats, with their enthusiastic grooming regimes, are simply less smelly than dogs, and therefore less easy to identify. Our pet cats have evidently worked out that we can't smell their smells and have resorted to more tactile and visual methods of capturing our attention, thereby further discouraging us from using our noses when we interact with them.

Pungent Plants

Cats are probably one of the strictest carnivores on the planet, so much so that they are often referred to as hypercarnivores or obligate carnivores. Their lives are all about meat. And yet they have a strange penchant for certain plants. Not to eat as part of their diet—more as an occasional interactive indulgence.

There is a very old book, published in 1768, written by botanist Philip Miller and titled *The Gardeners Dictionary*. A comprehensive description of all things botanical and garden related, it includes this brief passage about the plant *Nepeta cataria*: "It is called Catmint, because the cats are very fond of it, especially when it is withered, for then they will roll themselves on it, and tear it to pieces, chewing it in their mouths with great pleasure."

This is thought to be one of the very earliest written descriptions of the effects of this plant on the domestic cat. The *N. cataria* form of catmint, a perennial herb that produces small white flowers, is more often referred to as catnip nowadays, to distinguish it from other varieties of catmint that lack the seductive effect on cats.

Miller's short description briefly touches on a much more extensive behavioral display that a mere waft of catnip produces in many cats. Although cats sometimes perform the flehmen response when they sniff catnip, research revealed that this particular smell is processed via the cat's normal nasal olfactory system rather than through the vomeronasal organ. After sniffing, they typically rub their faces on the source of the catnip, often licking or sucking it and drooling in the process, rolling ecstatically and playing with the item. For the modern-day cat, the item, more often than not, will be a small soft toy filled with dried catnip. This dried version of catnip has been a mainstay of the cat toy industry for many years, its inclusion in products designed to entice cats to play. The apparently euphoric reaction to catnip usually lasts around ten to fifteen minutes and then wears off, followed by a period of up to an hour when the herb no longer has the same effect. This waning is only a temporary feature, however—as owners of catnip-loving cats will know, the same catnip toy will be just as much fun the next time it is hooked out from under the couch.

Such surprising behavior in cats inevitably drew the attention of scientists keen to find out exactly what the story was with this intriguing plant. Geneticist Neil Todd in 1967 discovered that the response to catnip is inherited via a dominant gene, with only two-thirds of cats exhibiting any reaction to it. Kittens, even with the appropriate catnip gene, don't begin to respond until at least three months old and often not till six months. The secret ingredient of catnip turns out to be a compound known as nepetalactone, and the response to it is widespread throughout the Felidae family, with many of the bigger cats such as tigers, ocelots, and lions exhibiting similar behavioral routines on smelling it. Quite why nepetalactone produces this response in cats and their relatives, though, remains a puzzle.

It's not just catnip—more plants have been discovered that have a special lure for cats. These include Tatarian honeysuckle, valerian root, and Japanese silver vine. Scientists looking at silver vine found that the component that cats react to in this plant is nepetalactol, which is similar to the nepetalactone in catnip.

In a fairly overlooked paper from 1964 titled "Catnip: Its Raison d'Être," Thomas Eisner from Cornell University described a

series of experiments showing that nepetalactone has the ability to repel certain insects. He postulated that nepetalactone serves a protective function, deterring plant-eating insects from consuming plants, such as catnip, that produce it. With Eisner's discovery in mind, the silver vine researchers decided to explore this effect with cats. They discovered that cats that had rubbed their heads and faces on the leaves of silver vine benefited from a repellent effect on mosquitoes. That the crazy behavioral response of cats to these plants might actually have a practical benefit, albeit probably an accidental one, is something that doubtless will receive a lot more attention.

It was eight weeks later. For Sheba, as with most domestic cat mothers, the joy of motherhood was beginning to wear thin as her six crazy kittens raced and tumbled around her. Occasionally, when she lay down, one would try an opportunistic suckle from her and she would rapidly shake them off. The kittens were fully weaned now, almost independent from her, and I watched them wistfully as I put down their saucers of food and they ran up, excited by the scent of their meaty meal. Soon they would all move to new homes, each taking with them a small piece of the blanket from their sleeping box, full of smells from their mother and siblings. It would help them adjust to the enormous changes and challenges ahead—strange smells, different households, and learning to communicate with their new owners.

CHAPTER 3

YOU HAD ME AT MEOW

Cats seem to go on the principle that it never does any harm to ask for what you want.

—Joseph Wood Krutch

"That one in there—he just sits and hisses." The school caretaker pointed to a hole underneath the old building. I crouched down, peered in, and said, "Hello there," to the dirty, scrawny little cat, who promptly hissed at me with all his tiny might. Hissing Sid, as he became affectionately known, was one of a colony of feral cats that my colleagues and I went on to rescue from the grounds of the school, where they were becoming something of a nuisance. After a little sojourn in a rescue shelter where they were all neutered and their kittens found new homes, the cats were relocated to a farm. Over the next few years, feeding them in their special cat shed on the farm every day, these cats became part of my life. Here, as they learned to trust me, they worked out new ways to communicate with me. Ways that included less hissing and more of the friendly sounds we associate with our sweet-talking pet cats.

In a cat's world, where smells are paramount, it must be a bewildering experience when they first hear a person speak. So many different, unfamiliar sounds directed either at another person or, even more perplexingly, at the cat. Humans are very preoccupied with the spoken word, babbling away at everyone and everything we meet. Intrigued as to what their "spoken" sounds mean, we have developed something of a fascination with the vocalizations of cats too. Nestled deep in the history books, a diary entry by the Abbé Galiani of Naples, dated March 21, 1772, offers some of the earliest recorded insights into cat vocalizations.

"I am rearing two cats and studying their habits—a completely new field of scientific observation . . . Mine are a male and a female; I have isolated them from other cats in the neighborhood, and have been watching them closely. Would you believe it—during the months of their *amours* they haven't miaowed once: thus one learns that miaowing isn't their love language, but rather a signal to the absent."

Little did he know it, but Galiani was ahead of the game with his observation that his two cats never meowed to each other. The true purpose of meowing would only be discovered centuries later, when larger scientific studies of cats became more accepted.

Through the intervening years, feline literature embarked on something of a magical mystery tour of the apparent linguistic talents of cats. Writers mostly attempted to define cat vocalizations along the lines of human language, identifying consonants and vowel patterns and certain "human" letters in their cats' speech. Reflecting on the differences between cats and dogs, Dupont de Nemours, an eighteenth-century naturalist, wrote, "The cat, also, has the advantage of a language which has the same vowels as pronounced by the dog, and with six consonants in addition, m, n, g, h, v, and f."

Some authors took this a step further to describe cats' use of actual human words. In 1895 Marvin R. Clark, a musician and lover of cats, published an enchanting and slightly bewildering book titled *Pussy and Her Language*. In this he includes "A Paper on the Wonderful Discovery of the Cat Language," apparently penned by a French professor named Alphonse Leon Grimaldi. In it, Grimaldi claimed to have elucidated the language of cats, providing an in-depth analysis of the cat's use of vowels, consonants (apparently used "daintily" by cats), and grammar, as well as words and numbers.

Grimaldi's paper included a list of what he considered to be seventeen of the most important words in the feline language:

GRIMALDI'S FELINE DICTIONARY	
Aelio	Food
Lae	Milk
Parriere	Open
Aliloo	Water
Bl	Meat
Ptlee-bl	Mouse meat
Bleeme-bl	Cooked meat
Pad	Foot
Leo	Head
Pro	Nail or Claw
Tut	Limb
Papoo	Body
Oolie	Fur
Mi-ouw	Beware
Purrieu	Satisfaction or Content
Yow	Extermination
Mieouw	Here

He went on to elaborate, "In the feline language the rule is to place the noun or the verb first in the sentence, thus preparing the mind of the hearer for what is to follow." As if this weren't skilled enough, Grimaldi also considered cats capable of counting. He compiled a comprehensive list, including "Aim" for number one and "Zule" for millions.

Grimaldi's "translations" were not surprisingly met with mixed reactions; many authors dismissed them as nonsense. However, among his rather bizarre suggestions, he did include a few wonderful nuggets of insight. His description of an enraged cat, for example, will resonate with many people:

"The word 'yew' . . . when uttered as an explosive, is the Cat's strongest expression of hatred, and a declaration of war."

In 1944, Mildred Moelk revolutionized the world of cat language with her in-depth study of the phonetics of the sounds produced by her own house cats. Her approach was to divide the vocal sounds of domestic cats into three main categories based on how they are produced. First, those made by the cat with their mouth closed, such as purrs, trills, chirrups, and murmurs. Second, the sounds made while the cat's mouth is opened and then gradually closed—these include the meow, the male and female mating calls, and the aggressive howl. The last group are all made while the mouth is held continuously open, generally associated with aggression, defense, or pain in cats. They include growls, snarls, yowls, hisses, spits, more intense mating cries, and shrieks of pain.

The difficulty in this vocal categorization lies in the huge amount of variation in the production of sounds, both between cats and within the repertoire of a single individual. As Moelk so elegantly put it, "The house-cat, unlike man, has enforced upon it no model of traditional language and no standard of correct pronunciation to which it must conform." Her work has been used as the basis for the analysis of cat vocalizations ever since. Some investigators have attempted to classify them using phonetic criteria like Moelk, while others have examined their acoustic qualities or concentrated on their behavioral contexts.

Although cats have a huge range of vocalizations, in cat-to-cat interactions they generally reserve these sounds for three types of occasions: finding a mate, fighting, and communicating between kittens and their mothers. The first two involve supernoisy sounds that we tend to hear at nighttime. Caterwauling, shrieking, bloodcurdling noises—the sorts of calls that make you rush outside to identify the source or cover your ears to block them out. In their quest to communicate with humans, cats seem to have ingeniously worked out that it is the gentle sounds, like those used between a mother cat and her kittens, that appeal to us most.

The Cat's Meow

Newborn kittens start life with the ability to purr, spit, and produce a few simple "mew" noises. At least they sound simple to us. What sounds like a lot of squeaking to the human ear is actually a range of different kitten calls. In addition to crying when they are hungry, kittens have a distress call that varies in tone, length, and volume depending on the reason for their anxiety. The mew of a kitten that is too cold has the highest pitch; becoming lost from the nest produces the loudest mew; and the most urgent and persistent mew is reserved for when they are somehow trapped. This last cry often happens as the mother sprawls out on her side to allow her kittens to nurse, inadvertently lying on some of them in the process. Depending on the type of cry, she responds by retrieving the lost kitten or by changing her position a little. Shifting her body as she lies nursing her litter encourages a kitten that has dropped off a nipple and become chilled to snuggle back in, or enables a squashed kitten to wriggle back out.

A study by Wiebke Konerding and co-researchers looked more closely at the responses of both male and female adult cats to recordings of two different types of cries made by kittens. One type had been recorded in what the authors describe as a "low arousal" context, made by kittens that had simply been spatially separated from their mother and the nest. The other was recorded in a "high arousal" context in which, as well as being separated from their mother, the kittens were held by the experimenter (restrained/trapped). On hearing these recordings, adult female cats oriented themselves toward the source of the cry (a loudspeaker) faster for the more urgent (trapped) kitten calls compared with the less urgent (strayed from nest) ones, indicating that they distinguished between the two. This happened regardless of whether they had ever had kittens themselves. Male cats, on the other hand, although they reacted to the kitten cries, showed no difference in their reactions to the two call types. Female cats therefore seem somehow hardwired to identify distress calls of kittens. Studies have also shown that each kitten develops its own individual versions of these calls and that these remain constant as it grows older. Whether mother cats can recognize their individual kittens from their calls alone remains unknown.

In turn, mother cats have a very special type of call they use when interacting with their kittens. Often described as a chirrup or chirp, this gentle trill-like sound was written by Moelk as "mhrn"* phonetically. It is a delicate, cheerful sound, described by the nineteenth-century writer Lafcadio Hearn as "a soft, trilling coo, a pure caress of tone."

To humans this enchanting call sounds much the same in all

---

*Think of a dove cooing "mhrn, mhrn, mhrn" with a trill on the *r*.

mother cats. Kittens, though, can recognize the chirrup of their own mother when they are only four weeks old. They can distinguish it not only from her meows but also from the chirrups and meows of different mothers. Researchers discovered this by videoing and analyzing the responses of four-week-old kitten litters when hearing vocalizations of their own and other mother cats. While a mother cat was absent from the room, experimenters played recordings of vocalizations from behind a screen to her litter of kittens still in their nest. They played them a meow and a greeting chirrup from their own mother as well as a meow and chirrup from an unknown mother cat, at an equivalent stage of motherhood to their own. Looking at the kittens' responses, the researchers found that they became alert faster to chirrups than to meows. They also stayed alert longer, were quicker to approach the source of the sound (the loudspeaker), and stayed there significantly longer when hearing their own mother's chirrup compared with any other of the sounds. That kittens can do this from such an early age suggests an advanced level of cognition at a time when they are only just beginning to move around and explore their world. This may be an adaptation for survival in the wild, where litters of kittens are often hidden out of sight by their mother while she goes off to hunt or find food. Her reassuring chirrup as she returns lets them know that it is safe to come out.

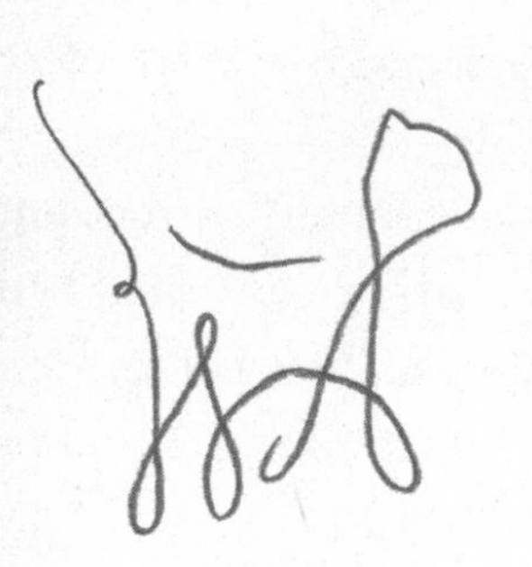

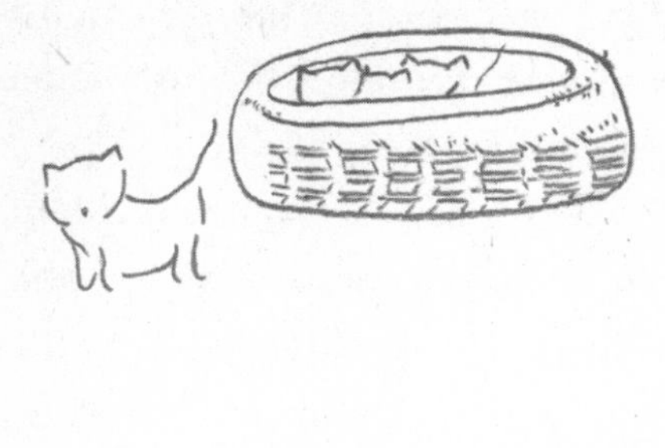

As kittens mature into adult cats and their vocal cords develop, their tiny mews gradually change into the more elaborate sounds that we describe as "meows." I'd been studying my adult hospital and farm cats for a while before I realized, just like Galiani back in 1772, that I had never heard them meow to each other. They would hiss occasionally and may well have quietly purred when sitting together, but that was the extent of their vocalizations. Later studies confirmed this discovery—the iconic meow of adult cats is almost exclusively reserved for cat-human interactions.

In the wild, away from the comforts of a human home, mew vocalizations gradually decrease as kittens become more independent. In house cats, though, meows are by far the most frequent vocalizations directed toward humans. Our pet cats often combine the meow with extra sounds such as trills or purrs. Some cats, like people, are chattier than others. Certain pure breeds, particularly oriental ones such as Burmese and Siamese, have a reputation for being more vocal. That said, many random-bred house cats, or moggies, spend their days meowing hopefully at their owners.

So why do they meow at us? It seems that over the ten thousand odd years that they have associated with us, cats have learned that we don't always understand their wonderfully subtle language of scents, twitches of the tail, and flicks of the ears. They need to make noise in order to get our attention. And lots of it. For the ever-adaptable cat, what could be more logical than to use vocalizations that, as a kitten, so effectively achieved a response from their mother?

What exactly is a meow? A simple answer is hard to find, and it depends on who you ask. Nicholas Nicastro from Cornell University has studied the meow and our understanding of it extensively. His wonderful though head-spinningly technical definition describes the acoustics of the meow:

> . . . a quasiperiodic sound with at least one band of tonal energy enhanced by the resonant properties of the vocal tract. The call ranges between a fraction of a second to several seconds in duration. The pitch profile is generally arched, with resonance changes often reflected in formant shifts that give the call a diphthong-like vowel quality. . . . This call type very often includes atonal features and garnishments (trills or growls) that may serve to differentiate the calls perceptually.

A slightly simpler, more phonetic version comes from Susanne Schötz and her team in the Meowsic project at Lund University in Sweden: ". . . a voiced sound generally produced with an opening-closing mouth and containing a combination of two or more vowel sounds (e.g. [eo] or [iau]) with an occasional initial [m] or [w] . . ."

Urban Dictionary's definition is far more succinct but to the point: "Meow is the sound a cat makes. It is also the sound a human makes when they are imitating a cat."

To the human ear, meows can sound friendly, demanding, sad, assertive, persuasive, persistent, plaintive, complaining, endearing, and even annoying. Some investigators have attempted to categorize meows into different subdivisions, but their classification proves tricky because, just like other cat vocalizations, the meow varies substantially among cats—and even changes in the same cat at different times. Despite this variability, there seems to be a word for "meow" in every language, from the Danish "mjav" to the Japanese "nya."

However we choose to say or spell it, the sound of a cat meowing is unmistakable. Unless that meow you thought you heard is actually a baby crying? Both sounds are generated by the vibration of the vocal cords in the larynx, and the acoustics of the two are remarkably similar, particularly with respect to what is known as fundamental frequency, or the number of waves of sound that occur per second. To the listener this frequency is perceived as the pitch of the sound—the higher the frequency, the higher the pitch. The cries of healthy babies have been shown in various studies to have an average frequency of 400 to 600 Hz and are described as having a falling or rising-falling pattern as the cry continues. Adult domestic pet cat meows, although hugely variable, were found by Nicastro to average 609 Hz. Other researchers, such as Schötz, have reported similar figures.

Pitched around the same level, both cat meows and baby cries seem to be particularly hard to ignore. The much-researched cries of babies have been shown to elicit alertness and distress in adults. In fact, Joanna Dudek and coworkers from the University of Toronto demonstrated that hearing babies' cries affects our ability to perform other tasks. No one has tested yet whether cat meows have the same effect but, given the acoustic resemblance to baby cries and the creativeness of cats, we can probably assume they are quite distracting.

Is this why cats are so hard to ignore? Have they somehow hot-wired our brains so we simply must respond to an urgent need to take care of them like a baby? Possibly yes, but probably not intentionally. Throughout domestication, we may have unwittingly selected for cats with the most persuasive meows, those that tend to resemble the cries of our own infants. Nicastro's study showed that compared with African wildcats (the ancestors of the domestic cat), the meows of domestic pet cats sound much more pleasant to human listeners. This may well be related to the differing pitches of their vocalizations, with the wildcat calls averaging 255 Hz compared with the much higher 609 Hz pitch of the domestic cats. Another study, exploring the acoustics of feral cat and pet cat meows, found the pitches of feral cat meows to be much lower than those of pet cats too. The meows of the ferals more closely resembled those of the wildcats in Nicastro's study. This suggests that socialization and experience with humans in some way modifies the meows of domestic cats.

Interestingly, while feral cats barely meow at all when first looked after by a human, rescue workers often report that ferals increase their rate of meowing as they spend more time in their company. Even some of the feral cats that I watched on the farm, who only ever came near me very briefly when I dished up their food before leaving each day, gradually began to learn to meow a little as time passed. Cats learn fast.

What appears to be a very deliberate adaptation of normal vocalizations to manipulate other species has been found in a wild species of cat, the margay (*Leopardus wiedii*). This was revealed in the findings of a scientific study by Fabiano de Oliveira Calleia and co-researchers conducted in the Brazilian region of the Amazon rainforest. Looking for as much background information as they could find on the various wildcat species in this area, the researchers interviewed local people who lived in the jungle. From these interviews emerged anecdotal reports of cougars (*Puma concolor*), jaguars (*Panthera onca*), and ocelots (*Leopardus pardalis*) all mimicking the vocalizations of their prey (such as agoutis and birds) in an attempt to attract them.

One day the researchers themselves were studying a group of pied tamarins, a small primate that they had been tracking, when they witnessed a margay attract the attention of the lookout tamarin, or "sentinel," by imitating the call of a tamarin pup. The confused sentinel climbed up and down the tree checking out the infant sound and alerted the rest of his group to it. This strategy of the margay, although unsuccessful on the day Calleia and coworkers observed it, presumably allows the cat to lure prey into a more accessible position for attack by masquerading as one of them. The researchers point out that the prey species mentioned by the Amazon locals in their interviews all produce vocalizations with characteristics that might be reproducible by cats. Choosing to mimic the call of the prey's young is particularly ingenious—as with humans, it pretty much guarantees a reaction.

Wild animal species tend on the whole to be fairly quiet, part of an inbuilt fear response essential for their survival. While some sounds are necessary for attracting mates, seeing off enemies, or raising alarms, for many, communication via scent or visual signals is a much safer bet for avoiding detection by predators. Similarly, for predators like the African wildcat ancestor of the domestic cat, the silent approach is far more likely to end in a successful hunt and a meal. Chitchat is, in fact, a luxury to be enjoyed only by species such as our own, those that don't have to worry about being chased and eaten, or about chasing to eat.

The process of domestication favors animals that are less fearful of people, those that best respond to being around us. Along with this relaxation of fear comes a reduced tendency to be silent, and so vocalizing in new and different situations becomes an option. As Darwin himself put it, "We know that some animals, after being domesticated, have acquired the habit of uttering sounds which were not natural to them." Take dogs, for example. Their ancestral wild species, the wolf, barks as a puppy but much more rarely as an adult, by which time barks comprise only around 2 percent of their vocalizations. Juvenile wolves are discouraged from barking as they mature to avoid attracting unwanted attention or scaring away prey. Yet, as anyone living in a residential area of a city will know, many adult domestic dogs bark a lot. On those rare occasions when adult wolves do bark, it is for only two reasons—territorial defense or to alert others. Dogs, on the other hand, are not only more "barky" generally but also use their barks in many different circumstances.

What Are You Talking About?

> *When I meow it means . . . I am hungry . . . I want food in my bowl . . . I want food in my bowl right now . . . I want to go out . . . I want to come in . . . Brush me . . . Get my toy out from under the sofa . . . It's time to change the litter . . . I just put a mouse in the bureau drawer . . . I did not break that vase . . . Get me down from this tree . . . Please kill that dog next door . . . Hello . . . Good-bye.*
>
> —Henry Beard

When we listen to cats, can we tell what they want from those meows that they so skillfully direct at us? Many people feel they understand what cats are saying when hearing them meow. However, under the scrutiny of scientific tests, it seems we may rely on more than just their meows to get the full picture. In a pioneering experiment, Nicholas Nicastro recorded the meows of cats in five different situations: requesting food, annoyed at being brushed, requesting attention, requesting to be let out, and distressed when traveling in a car. He played them back to people and discovered that, although they performed better than chance at identifying the meow contexts, without the added visual cues gained when seeing the vocalizing cat, people found this quite difficult. This conclusion was also reached by subsequent similar studies.

This suggests that cat meows simply grab a person's attention and convey the need or desire for *something* rather than actually conveying detailed information to humans about what they want. As Nicastro put it, cats may meow "to provoke, rather than to specify, a reaction." Once cats have used the meow to gain a person's attention, they generally use additional visual or tactile techniques for explaining what is so urgently needed, such as rubbing their head and flanks around our legs and then against the food cupboard, or sitting looking pointedly at the back door.

For years scientists thought the same about dog barks. Gradually, though, researchers realized that the acoustic structure of barks varied according to the context in which they were made. For example, barks in response to the doorbell ringing are harsher, lower pitched, longer, and more repetitive than those made when dogs are playing or left alone.

So, are cat meows really just meaningless attention-getting sounds? Just like kittens with their squeaky mews, adult cats appear to have their own individual repertoires of meows. This is partly what makes it hard for people to tell the contexts apart. Some scientists have suggested that these variations might be similar to the different dialects and languages of people. Nicastro's study dug a little deeper and discovered that, as with dog barks, the picture is not quite so straightforward. He found that experience with cats improves people's ability to identify meow contexts. A later study by Sarah Ellis and co-researchers also showed that cat owners are more successful at distinguishing meows when listening to recordings of their own cats rather than those of unfamiliar ones. Other researchers have found that female participants proved better at identifying different types of meow contexts, possibly related to their higher scores on level of empathy toward cats.

All in all, the consensus is that identifying meows is hard, but not impossible. Over time, pet cats may learn to vary their meows as their owners become tuned in to their cat's individual sounds and can recognize the different meanings. Nicastro cites this as an example of "ontogenetic ritualization," a process during which members of two species gradually shape each other's behavior by repeating a social interaction.

If humans can learn to decipher meows as they gain experience with cats, then it seems likely that some real information is contained within this type of vocalization. Many scientists now agree that, while other species make different sounds to us, certain characteristics of vocalizations are similar across many animal "languages." Tamás Faragó and his coworkers showed that when listening to dog vocal expressions, humans use the same simple innate acoustic rules that they use for assessing the emotional content of another person's vocalizations. This taps into a concept first mooted by Darwin in 1872: "That the pitch of the voice bears some relation to certain states of feeling is tolerably clear." Processing of emotional content is therefore probably something that we naturally do without realizing it when listening to vocalizations from any other species, including cats.

Researchers have looked in more detail at the acoustic qualities of cat meows. One study found that meows recorded in a pleasant situation (receiving a snack) had higher average pitches compared with those recorded in an unpleasant situation (inside a cat carrier in a car). In a different study, Susanne Schötz and her team in Sweden also found that consistent subtle differences between meows in different contexts exist. They discovered that how cats are feeling affects both the pitch and the way it changes during the meow. So a positive meow (e.g., in greeting or to request food) has a pitch that rises and ends up higher, while an unhappy or distressed meow (e.g., when traveling in a cat carrier) has a pitch that falls.

Human listeners may struggle to distinguish the subtleties between two positive meows recorded in different contexts, such as

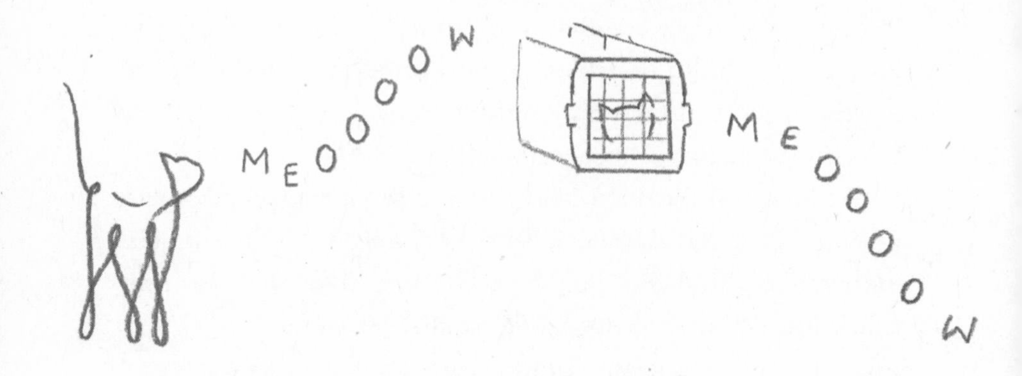

"requesting food" and "requesting attention." However, another study by Schötz showed that, when asked to tell the difference between positive/happy (food/greeting) and negative/sad (at the vet) meows, listeners performed significantly above chance. As in Nicastro's previous tests, cat-experienced listeners were better at this than those with no prior cat experience. It seems we can learn to pick out basic emotional information when hearing meows.

In a different study, Pascal Belin and coworkers, looking at our perception of positive (food-related and friendly) and negative (distressed) cat meows, made an intriguing discovery. Using functional magnetic resonance imaging, they monitored listeners' brain activity while they were played recordings of the meows. They found that even when listeners couldn't decide whether the meows were positive or negative, the sounds still registered differently within their brains. Negative vocalizations stimulated a greater response in regions of the secondary auditory cortex, whereas positive vocalizations stimulated greater responses in parts of the lateral inferior prefrontal cortex. Perception, if not recognition, of these positive and negative emotions appears therefore to be inbuilt—an interesting disconnection between brain activation and conscious behavior.

We seem to be particularly tuned in to sad vocalizations. A

wide-ranging survey of people's impressions of their pets revealed that pet-owning adults find distressed animal vocalizations sadder than adults without pets do. And cat owners are particularly sensitive to distressed cat vocalizations. This sensitivity to "sadness" in cats may be the price cat owners pay for the development of a better understanding of feline vocalizations. This is not necessarily a bad quality—from a welfare perspective, it is important that we are tuned in as much as possible to any sadness in our cats, as they are notoriously good at hiding illness and stress.

There is one particular cat "vocalization" that people often describe as sad—the voiceless meow. So incredibly effective, it was the inspiration for Paul Gallico's amusing and charming book *The Silent Miaow.* Cats seem to reserve this one for moments when they most need to tug at their human's heartstrings. Requiring the cat to have already caught the person's eye, the silent meow involves maintaining pleading eye contact while performing a soundless meow. Gallico's playful advice to cats in his book reads: "Don't overdo it but save it for the right moment." Cats are experts at this. A variation is what I describe as the "I can barely make any noise except this faint croaky squeak" version, which is almost, if not equally, as effective.

"But How Would You Ad-dress a Cat?"*

People love to talk. An observation evidently not lost on cats, given how creatively they have morphed their kitten calls into human-directed meows. People love to talk to their cats too. Many owners converse with their cats throughout the day, chatting away as if to another person. In one survey, 96 percent of owners

*T. S. Eliot, "The Ad-dressing of Cats," 1939.

reported that they spoke to their cat every day, and 100 percent spoke to their cat sometimes. Most of those asked quite happily admitted that they confide in their cats, talking to them about problems and important events. When they have been out of the house, owners returning from a longer period of time away talk more to their cats than when absent for a shorter time, as they might with another person.

Just as house cats have modified their meows to a sweeter-sounding higher pitch when talking to us, many owners change the tone of their voices and the way they speak when conversing with their cats. Most people when questioned say they talk to their cat as if they were a person, usually a child. The result is reminiscent of the special way we talk to babies and young children. "Motherese," as it is known, has been much studied as a concept, mostly with respect to human babies. It is seen, or rather heard, across many languages and cultures, and spoken by both men and women, not just mothers. Such talk tends to have a higher and greater range of pitch than normal speech, is performed more slowly, and contains lots of repetition. When directed at babies, motherese may also consist of simplified and exaggerated elements of language such as elongated vowels: "How are yooooou?" Research has shown that pet-directed speech tends to lack the elongated vowels but otherwise bears striking similarities to motherese.

Quite why we speak to our babies and cats like this is something of a mystery, although evidently not a recent one. In his bold and intriguing work published in 1897, *The Language Used in Talking to Domestic Animals*, H. Carrington Bolton describes it somewhat scathingly:

> Feeling a difficulty in making himself understood, man tries to lower his language to the level of animal

> intelligence somewhat in the same way that young mothers resort to that preposterous travesty of speech known as "baby-talk." Why infants and domestic animals are supposed to understand inarticulate sounds better than ordinary speech is difficult to explain; perhaps, however, as Bossuet wrote: "Les oreilles sont flattées par la cadence et l'arrangement des paroles." [Roughly translated as "The ears are soothed by the rhythm and tone of the words."]

Bolton's reference to Bossuet's rather enchanting words was close to the truth. Research has found that babies show a preference for baby talk over normal adult speech. The distinctive style of motherese gives it a "happy" rhythm and tone, which babies are particularly sensitive to. Baby talk likely contributes to their learning of language and helps create an emotional bond with the speaker. In talking to cats in a similar fashion, we may be subconsciously treating them like our own young. Or possibly we are unwittingly mirroring their meows with a similar high-pitched vocalization. Some people take this a step further and actually imitate the vocalizations of the cat they are interacting with. This tends to be mainly younger people who are actively playing with cats. An unusual practice, apparently particular to cat owners, and not generally reported for owners of other domestic pets.

While it is unclear what, if anything, cats take from motherese-style talk, it does seem to get their attention, at least some of the time. Many scholars studying speech have surmised that altering our vocalizations in this way is necessary for animals to realize that we are speaking to them. One small study looking specifically at the reactions of indoor cats to different types of human speech found that they could tell the difference between speech aimed deliberately at them and the normal speech that people use with adults. However, this was only the case when the speaker was their owner—they could not tell the difference between the two types of speech when it was a stranger doing the talking. This highlights again the significance of the gradual ritualization of communication in cat-owner relationships, especially for indoor pet cats, who often experience less exposure to strangers.

We don't need to adjust our pitch for cats to actually *hear* us. Cats' hearing range is one of the widest tested in mammals, covering 10.5 octaves compared to our 9.3. They can hear similar low-pitch sounds to humans but they have an upper range way beyond ours, enabling them to hear the calls of prey such as mice and rats. Then throw in their amazingly mobile ears that can rotate independently of each other, a full 180 degrees each, enabling them to locate sounds extremely accurately. They really have no excuse for ignoring us.

Cats often give a very good impression of not hearing a thing when we call, yet on closer inspection it appears they are really quite discerning of human voices. Researchers use the habituation/dishabituation technique to explore cats' reactions to different sounds. One such experiment, by researchers Atsuko Saito and Kazutaka Shinozuka, looked at cats' responses to different people calling their name. In a test resembling a kind of auditory identification parade, the cats listened to recordings of three strangers calling their name one by one, each separated by thirty seconds, followed by the owner calling their name. The strangers were all the same sex as the owner and were asked to call the cat's name in the same way as the owner did to make them sound as similar as possible. Analysis of the videos made of the cats while listening showed that their reactions declined with successive presentations of different strangers calling their name (habituation). However, when they heard their owner call, their response increased again (dishabituation), indicating that they recognized their owner's voice. They responded not so much with any reciprocal vocalization or obvious physical communication, but more through subtle movements of their ears and heads that increased when they heard their owner speaking instead of a stranger.

Following on from this study, and once again using the habituation/dishabituation technique, Saito and co-researchers investigated cats' ability to distinguish their own names from four general nouns that were similar in construction and pronunciation when spoken. They found that cats, having gradually paid less attention to the normal words as they were played, showed increased response again on hearing their name spoken. That they can pick out their own name in this way suggests a level of vocal understanding in cats that is generally associated with dogs and their eagerness to listen, understand, and please us—qualities rarely listed for cats.

Did You Have Something Else to Say?

While cats mostly rely on their meows to communicate with people, they have a few other sounds with which to charm us too. One of these is the trill. Reminiscent of the soft, gentle call used by mother cats to their kittens, this sound is often used by cats to greet people. It is sometimes heard as they approach their owner after a period away or in response to a greeting vocalization from a person. Cats often combine it with a meow to create a much longer sound. Most definitely a happy, friendly vocalization and one that, as Paul Gallico so accurately observes, has the following effect on people: "For some reason or other it just seems to make them feel fine, and puts them into a good humor."

This sound should not be confused with the unusual chittering or chattering that cats perform when watching an unreachable bird or other prey, often through a window or glass door. A strange teeth-chattering noise, this sometimes has vocal elements included too. Its purpose remains something of a mystery, most likely a frustration noise, although some researchers suggest that the cats are attempting to attract the attention of their prey, mimicking the chatter of birds, much like the ingenious margay of the Amazon.

Perhaps the most alluring sound cats make, however, is the purr. Mark Twain once said, "I simply can't resist a cat, particularly a purring one." For many years it was a mystery how cats even produced the sound, let alone what it actually meant. One early theory suggested that blood flowing through veins in the chest produced the purring sound. Scientists gradually realized that the sound emanates from the throat area. We now know that it is controlled by a neural oscillator, or "purring center," in the brain, which sends signals to the muscles of the larynx. These respond by rapidly opening and closing the space between the vocal cords known as the glottis, creating vibrations at a rate of 25 to 150 per second as the cat breathes in and out. The result is an almost continuous purring sound.

And it's not just the domestic cat that can purr—many of the larger wild felid species, such as cheetahs, can also make this impressive rumbling sound. Interestingly, big-cat species that purr can't roar, and those that roar, like the lion, can't purr. This is thought to be due, at least in part, to differences in the structure of the vocal cords, which are much larger and fleshier in roarers. Just like the meow, domestic cats appear to have their own individual purrs that vary in characteristics between the inhaling and exhaling phases.

Exactly why cats purr remains something of a mystery. Able to produce the sound from early kittenhood, domestic cats first purr nestled in among their mother's fur while suckling, alongside any siblings they have. Later in life, they may purr when in contact with humans or other cats in friendly situations; when they are sleepy; or when they are warm and cozy. One study recorded the behavior of cats when their owner had been out of the home for thirty minutes compared to four hours. The researchers found that the cats purred significantly more on their owner's return when they had been absent for a longer period.

As always with cats, the story is not quite that simple. Domestic cats purr in many more situations than when curled up contentedly on a human's lap. In apparent contradiction to its occurrence in these peaceful situations, cats may also purr when in far more stressful situations, such as when visiting the vet. A survey taken within one veterinary clinic showed that 18 percent of cats purred while being examined by the vet—most definitely not a warm and cozy situation for a cat. Some cats also purr when in pain and even when dying. Although a definitive explanation of purring has remained elusive, it may in some way be self-soothing.

Whatever its natural purpose, domestic cats have certainly turned purring to their advantage in communication with people. Some purrs contain a higher-pitched meow type of vocalization within them, so that they sound more musical in nature. These tend to occur when the cat is hungry and trying to persuade their owner to feed them. Karen Mccomb and her team from the University of Sussex discovered this "cry embedded within the purr" by analyzing the acoustics of purrs recorded in different situations. When listening to recorded purrs, people can distinguish between normal and meow-containing purrs, finding the latter more urgent or demanding in nature and less pleasant than a regular contented purr. The meow component of these urgent purrs, just as with normal meows, has a strong acoustic similarity to the cry of a baby, making these "food-soliciting" purrs difficult to ignore.

Where Next?

It's an impressive achievement. Domestic cats, opportunistic as ever, have learned to adapt their calls of kittenhood to tap into our brains and tug at our unsuspecting heartstrings, unwittingly (or maybe not) masquerading as one of our own young. In return, we, too, have adapted our speech, often talking to them as if they really were our babies. We roughly get the gist of whether they are happy or sad. But most of the time, once they've got our attention with a meow, they need to physically show us what they're actually talking about. With some work, though, they can train us up so we understand them better. Meanwhile, cats quite obviously understand far more of what we say than they are letting on, picking out the most important word (their name) from our constant chatter in hopes that it means something good is coming.

Is that it? Has cat-human vocal communication gotten as good as it will get, or is it still evolving? In evolutionary terms, cats and humans have had only a relatively short time to work on it—a mere ten thousand years since we started keeping each other company. Combine that with the fact that cats are an innately solitary species, undisposed to using their voices, and it is a miracle that we can communicate vocally at all. Cats are surely unlikely to leave it at that when they have so much to say for themselves.

The little black-and-white cat was sitting waiting for me just outside the shed door. Almost one year after his rescue from the school grounds, he had filled out a lot, and his coat was thick, healthy, and clean. Hissing Sid had by now become something of a poster boy, advocating the benefits of neutering with his "before" and "after" pictures featured in cat magazines. Avoiding eye contact as I had learned to do with him, I unlocked the shed and set about putting out dinner for him and the other colony cats. As Sid sat at my feet, I heard his rumbling, motorboat purr start up. I chatted away to him and suddenly he graced me with a plaintive meow. He had finally learned the art of talking to me—well, at dinnertime, at least. Thrilled, I turned to look at him, bent down, and gently held my hand out for him to sniff. Tentatively Sid sniffed. Then hissed. Old habits die hard.

CHAPTER 4

TALKATIVE TAILS AND EXPRESSIVE EARS

The language of the tail cannot be misinterpreted, suggestive as it is of the feelings of the Cat.

—Alphonse Grimaldi, 1895

It was almost midday, lunchtime for the small group of cats I was watching on the hospital grounds. The door at the top of the ramp leading into the kitchens opened and a tray of leftovers slid out; scrambled egg was on the menu today. Right on cue Frank, the resident large tomcat, sauntered back from his morning patrol. Betty, one of the female cats, approached him, speeding up as she went. As she trotted toward him, her tail lifted and pointed straight up to the sky. I made a note: "Tail Up."

Strolling through the hospital gardens later that day, I was mapping out locations of other cats (those that weren't part of my observation group) when one of the more friendly ones, Flo, came up to me. As I bent down to say hello, I noticed it again—that raised tail as she approached me before rubbing around my ankles.

Hardly a Eureka moment, but it did make me wonder as I mulled it over on my way home. Why did Flo raise her tail like that to me, just as Betty had to Frank? Did Flo regard me as another, albeit huge and two-legged, cat? If so, were we equals? I had seen young kittens running up to their mothers with their tiny tails pointing straight up, so maybe she thought of me as a slightly strange-looking mother figure. What did Tail Up mean to a cat? Was it a deliberate signal for something or just a subconscious action, like we might chew our lip when anxious or smile when we feel happy?

This, as well as their other behavior, fascinated me, and I expanded my research to a second group of feral cats on a nearby farm. I spent almost a year studying the interactions of each group, recording everything I could, including how they used their tails.

As animal species have evolved to inhabit different ecological niches, their tails have taken on many new looks and styles. There are enormous tails, seemingly out of all proportion to the rest of the body; prehensile tails; fluffy, feathered, spiny, and slinky tails; tiny vestigial tails; and in some cases, such as our own species, the complete absence of a tail.

Tail functions have become equally diverse. For organisms with no other limbs, such as fishes, the tail is on the must-have list, essential for locomotion. Even with the welcome addition of legs on which to move around, animals often still use their tails to aid balance and coordination. Squirrels, for example, use their fluffy tails to stabilize themselves as they jump from tree to tree. Researchers have found that kangaroos, when grazing and moving slowly, use their tails like an extra leg.

Some species have adapted their tails to become prehensile and hang on to nearby props. Included in this impressive lineup is the tiny harvest mouse, which uses its tail to climb grass stalks, and the ever-enterprising seahorse, which clings to strands of seaweed to take a rest from swimming. More substantial prehensile tails are seen in a range of New World primates—these tails support the entire body weight of monkeys as they swing from tree to tree to find new feeding sites. Such strength and flexibility have required considerable evolutionary changes in the structure of bone and muscle in the tail.

Tails for many animals have developed uses far beyond movement and balance. Some pangolins, porcupines, aardvarks, and lizards use theirs as weapons, most commonly against predators.

In animal species likely to be preyed upon by others, the tail is often used to make a signal in response to seeing a predator. Such displays may be warnings to their own species of an approaching predator or signals to the predator itself letting it know it has been detected and has lost the benefit of surprise. Or both—the California ground squirrel, for example, flags its tail when it spots a snake lying in wait, alerting other squirrels and often causing the snake to abandon both its current predatory mission and its now-outed hiding place.

One aspect of tail movement that particularly interests scientists is its potential to show how an animal is feeling. This has formed the basis of many studies hoping to improve animal welfare, particularly for domesticated animal species. Studies of dogs, for example, have revealed far more subtleties to the well-known tail wag than we might ever have suspected. In one such study, Angelo Quaranta and coworkers presented thirty dogs with four different stimuli—their owner, an unknown person, an unfamiliar cat, and an unfamiliar dominant dog—and recorded how the dogs wagged their tails in response to each. They discovered that when they were happy and excited to see their owner, the dogs directed their wag more to the right. A similar right-biased wag, although somewhat less enthusiastic, occurred when viewing a stranger. When eyeing up a cat, the wagging was much reduced but still showed a bias toward the right. Faced with the unfamiliar dog, however, the dogs wagged more to the left.

These patterns of left- or right-biased wagging may be due to different emotions eliciting responses in separate sides of the brain. The left side of the brain, which when activated produces a right-biased wag, is thought to be associated with approach-type responses and so may be stimulated by the dogs seeing their owner, another person, or a cat. Seeing an unknown dog, however, might elicit a tendency to withdraw, controlled by the right-hand side of the brain and causing a left-sided bias to the tail wag.

In order to test whether other dogs notice this asymmetrical wagging, Marcello Siniscalchi and co-researchers presented dogs with videos of other dogs wagging their tails with biases either to the left or to the right. They found that the observing dogs exhibited elevated heart rates and higher behavioral scores of stress and anxiety when they saw another dog wagging its tail with a bias to the left, compared with one wagging toward the right. In other words, they appeared to sense from its tail movements that the other dog was in a withdrawal mode, a useful skill for avoiding potentially dangerous situations.

Farm animals show similar subtle variation in their tail movements, depending on what they are doing and how they feel about it. Cows, for example, tend to keep their tails still when lining up; wag their tails toward their bodies when feeding; and wag their tails vigorously when in contact with a mechanical brush. A slightly more alarming tail discovery comes from pigs. Scientists found that within a pigpen, seeing increasing numbers of individuals begin to uncurl or tuck their tails is a reasonable indicator of an impending outbreak of cannibalistic tail biting between sty-mates.

Tall Tails

A domestic cat's tail has, depending on the breed, up to twenty-three highly mobile vertebrae, along with an impressive set of muscles and nerves. This combination enables cats to move their tails in almost any direction—up, down, and side to side—and at varying speeds. This remarkable flexibility was not lost on the ancient Greeks—they called the cat "ailouros," from *aiolos* (moving) plus *oura* (tail). Modern-day domestic cats have acquired, with a little help from artificial breeding, a huge variety of tail types, from long and skinny to large and fluffy, and kinked, bobbed, or curly.

Wild members of the Felidae family, while lacking the tail variations of their domestic cousins, still possess the same basic flexible anatomy, making their tails excellent balance aids. The success of cheetahs as high-speed predators has been partially attributed to the use of their tails to stabilize their rapid movement. Scientists working on the development of robots were intrigued by the possibility of artificially reproducing this impressive maneuverability and incorporating it into their designs. Taking measurements of real cheetah tails, Dr. Amir Patel realized that although they look quite thick and weighty, these tails are actually surprisingly lightweight and their volume is mainly due to large amounts of fur. By suspending cheetah tails in a wind tunnel and examining their movements, he discovered they have significant aerodynamic properties that redirect and stabilize the animals as they move. For domestic cats, just as with their wild cousins, the tail is a great help in balancing. Less for pursuing antelope across African plains and more for tiptoeing along narrow garden fences and household shelves, it's a precision instrument, nonetheless.

As felids belong to a predator rather than prey species, their tails have some functions different from those of the animals they might hunt. Leopards, lions, and domestic cats, when crouched stalking their prey, lightly twitch the tip of the tail from side to side. Researchers have suggested that the tail may sometimes act as a snake-charmer-type "lure" to grab the attention of the prey, thus distracting its attention from the cat's face and, more importantly, jaws. Or it may more simply be that the twitching is due to frustration or anticipation of a meal to come.

Perhaps one of the most innovative recorded uses of a big cat using its tail was described by E. W. Gudger in 1946. His paper records individual accounts from various locals and explorers in Amazonia who all witnessed and described a similar phenomenon while observing jaguars. These observations, between the years 1830 and 1946, came from a wide area covering the rivers of southern Brazil and up to the headwaters of the Amazon. On the lookout for a fish supper, a jaguar would find a place where a fruit-bearing tree overhung the river. It would crouch under it, stretched out along a log or a branch that leaned over the water. The intended prey were tambaquis or other fruit-eating fish that would habitually rise to the surface when they heard fruit falling from the trees above into the water. Dangling its tail down while lying on the log, the jaguar would gently tap the surface of the water with the end of its tail, mimicking the effect of falling fruit. As the fish swam up to investigate, the jaguar would scoop them out with its paw. Ingenious.

As with dogs, it is as an outlet for expressing emotions that the cat's tail comes into its own. Domestic cats show many of their feelings through their tail movements, changing their tails from one position to another with enviable grace and producing a huge variety of semaphore-like moves. Canon Henry Parry Liddon, a nineteenth-century churchman and great lover of cats, described the cat's tail as a "catometer"—reflecting their ever-changing moods. He was right, and through a combination of science and getting our hands scratched when we read those signs wrong, we now know the meaning of many tail positions in cats. My favorite, the one probably most familiar to people and the most studied, is the Tail Up behavior I first saw Betty perform to Frank and that Flo approached me with on the grounds of the hospital that long-ago day. The tail is held vertical, not fluffed up, often while the cat is moving toward someone or another cat. The very tip may be curled slightly and waft around in the air. Occasionally, the tail will quiver—reminiscent of the pose cats assume when spraying urine against a vertical surface. Thankfully, cats rarely get these situations confused.

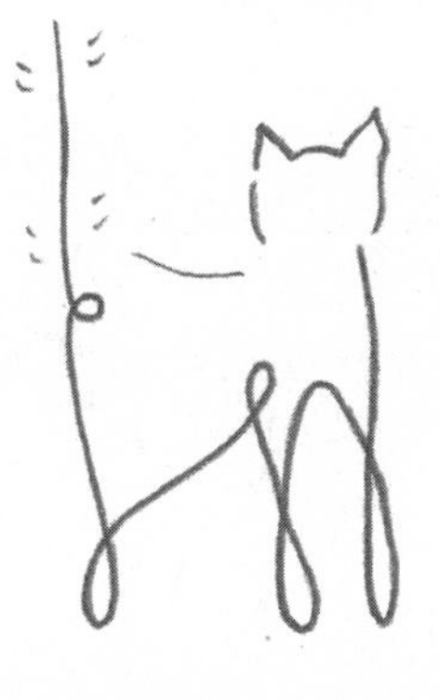

After many hours watching the hospital and farm colonies, I analyzed all the interactions I had recorded and a picture began to emerge of how the cats were using their tails to communicate. I discovered that when cats approached each other with their tails held straight up, there was little chance that any aggression would follow. In contrast, approaches with the tail lowered led to a more unpredictable outcome—sometimes the two cats would still appear friendly and simply sniff and then sit quietly with each other, and other times it would result in more hostile behavior. By performing a Tail Up, therefore, one cat seemed to be signaling to the other that they intended to interact amicably—a sort of "I come in peace" signal. The recipient cat often raised their tail in response, and then the pair frequently

Typical Interaction Involving Tail Up Between Two Cats

1. Dusty walks into the colony core area.

2. Penny raises her tail and approaches Dusty.

3. Dusty raises his tail in response.

4. Dusty and Penny exchange head rubs.

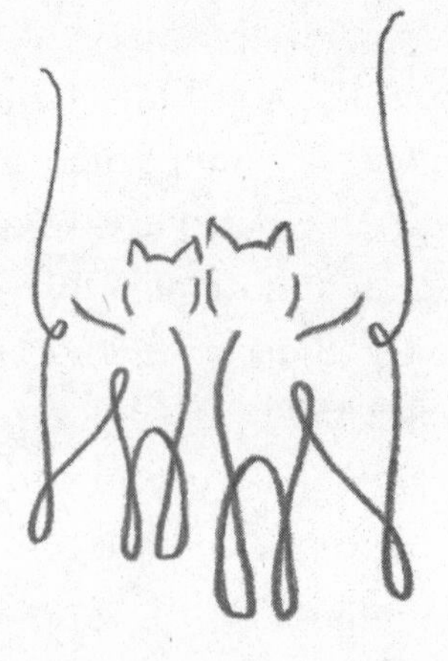

continued on to perform friendly behaviors such as rubbing their heads or bodies on each other (see opposite page). Sometimes, on receiving a tail-raised approach, the recipient would decide they weren't in the mood and the interaction would fizzle out, but the use of the Tail Up at the start generally ensured that the conversation was unlikely to turn nasty.

Following my early work, in a simple but elegant experiment, Charlotte Cameron-Beaumont at Southampton University tested the Tail Up behavior further by presenting domestic cats with cutout silhouettes of other cats, either with their tails raised vertically or sloping down toward the ground. On seeing the pretend tail-raised cat, the observing cats approached faster than when seeing an image with the tail down. They were also much more likely to raise their own tails in response to the Tail Up image. Sometimes the tail-down silhouette prompted tail swishing or tucking under of the tail in the observing cat, suggesting that they felt annoyed by or fearful of this tail-lowered impostor. This study confirmed that the raised tail is a signal recognized by cats as a friendly approach.

Other researchers began to wonder if the use of Tail Up within groups of cats followed any particular pattern—what exactly was the Tail Up etiquette? Italian researchers Simona Cafazzo and Eugenia Natoli looked at how the behavior was used within a colony of neutered cats in Rome. They ranked these cats according to the outcomes of aggressive encounters between different pairs of individuals and compared that to how they used their tail signals. There was a tendency for low-ranking cats to display Tail Up more frequently, whereas high-ranking individuals received it more frequently.

As is often the case with cats, though, the rules aren't quite that simple. Ranking cats on the basis of aggressive encounters often oversimplifies their relationships, especially in a large colony, where cats tend to avoid such encounters when possible. John Bradshaw looked at the Rome colony data from a different perspective. He noted that the females in the Rome colony rarely used Tail Up among themselves, even when their ranks, as determined by aggressive success, were quite far apart, and also that Tail Up was frequently observed being used from females to males. It was also used often by the juvenile male toward adults within the colony. Bradshaw concluded that, broadly speaking, Tail Up is a signal of friendly intent used by younger/smaller cats to those who are older/larger—kittens to mothers, juveniles to adults, females to males. How it is used within these categories is less clear—it may be more related to individual personalities and the history of the relationships between any particular pair of cats.

Of course, not all domestic cats live in outdoor colonies like these, scraping a living on farms, on the grounds of hospitals, or other places where concentrations of food handouts or leftovers might be found. Millions of domestic cats instead get to share our homes with us and sometimes with other cats too. Some of these pet cats may spend their whole lives indoors, while others are free to come and go and wander outdoors as far as they like. This can create varied challenges. Our cats may have to communicate with neighborhood cats outside. Inside, they may struggle to avoid other cats that live with them. Not only that, but there are humans to deal with too. For the pampered pet cat, the need for clear communication may sometimes be as great as for ferals in a crowded outdoor colony.

Most pet cats only have to share with one or two other house cats. One unusual study, however, focused on fourteen domestic cats that shared a single home together. As well as being a much larger group than ordinarily found living in a domestic environment, these fourteen cats were confined to the indoors. Despite being at a density fifty times greater than the cats in many outdoor colonies, aggression levels were much lower than might have been expected in such a confined environment. Observations suggested that tail signals were instrumental in reducing aggression between housemates. The authors, Penny Bernstein and Mickie Strack, suggest this worked "by 'tagging' individuals as being more or less likely to interact and/or be aggressive. Since the tail can be seen at a distance, receivers could tailor their responses before contact was imminent."

As demonstrated by Flo approaching me that day at the hospital, cats use the Tail Up behavior when interacting with humans too. They use the signal in much the same way they would with another cat—raising it as soon as they are within sight of the intended person. There is an advantage for cats when using tail signals in human interactions. Unlike with the cat's more subtle scent signals, there is a passing chance that a person will actually notice the tail signal at some point. A little feline vocal nudge in the form of a meow is often added to catch the human's attention. Sometimes, too, an enthusiastic tail quiver may get thrown in for good measure. Use of Tail Up varies under different circumstances, increasing noticeably in pet cats waiting to be fed. Possibly house cats use it in this context just as kittens might do to their mothers—to solicit food.

This isn't the only occasion that cats increase the use of this signal, though. As part of my doctoral studies, I carried out some experiments that looked at how cats use Tail Up and rubbing (which will be explored more in chapter five) when interacting with a familiar person, away from any situation involving food. Two different conditions were tested. In one a cat was released into a room, in the middle of which stood a familiar person who did not interact with the cat in any way (the "no contact" condition). In the other treatment, the same procedure was followed but the person petted the cat for twenty seconds per minute and talked freely ("contact" condition). In both cases, activity of the cat and the person was recorded on video for a period of five minutes and later analyzed to record the details of any interactions that took place. The results showed that the cats had their tails held vertical for a significantly higher proportion of the time when the person petted and spoke to them than when they ignored them. It seems that keeping the tail raised during reciprocal interactions with people is important for cats, perhaps showing that they feel the need to reinforce their friendly intentions as the exchange proceeds.

Pet cats tend to follow raised-tail approaches with a rub, just as they would with a cat. Often, they rub on the owner's legs and then entwine their tail around the legs. This Tail Up routine with a familiar person occurs far more often in cats that have been well socialized to humans as young kittens. It is also seen more in cats that have a natural tendency to be bold, a trait they inherit from their fathers, a topic explored in more detail in chapter seven.

An Evolutionary Tail

One of the most interesting aspects of the raised-tail greeting signal is that, across the forty-one different species within the cat family (forty wild species plus the domestic cat), it appears to exist only in the domestic cat (*Felis catus*) and the lion (*Panthera leo*). Observations of other wild felids have shown no evidence of it. Comparative studies have also been made of captive populations of Geoffroy's cat (*Oncifelis geoffroyi*), the caracal (*Caracal caracal*), the jungle cat (*Felis chaus*), and the Asiatic or Indian Desert cat

(*Felis lybica ornata*). The first two are from completely different branches of the cat family than the domestic cat, while the latter two, from the same *Felis* lineage, are more closely related to it. Although these species perform many of the behaviors shown by their domestic cousins, Tail Up is missing from all their repertoires. Even African wildcats (*Felis lybica lybica*), the ancestral species of the domestic cat, do not appear to exhibit Tail Up as adults, although their kittens perform the behavior to their mothers just as domestic kittens do to theirs.

Why do adult domestic cats raise their tails like this when nearly all other cat species don't? Most researchers agree that it is most likely due to the transition from a purely solitary existence in the wildcat to the flexibly sociable arrangement of domesticated cats today. The lion is the only other felid to have developed social living, and it, too, shows Tail Up. This suggests that Tail Up has evolved as a social signal separately in the domestic cat and lion, an evolution driven by the necessity of living in groups rather than by domestication itself.

The African wildcat would originally have had little need for visual signals for use with other cats. Other than during encounters for mating and interactions between a mother and her kittens, they would largely have communicated via olfactory means. Leaving long-lasting scent signals for other cats to read as they passed by would have been a far safer, remote method of making their point.

Courtesy of humankind some ten thousand years ago, these wildcats began to congregate around larger sources of food and needed to find a way of avoiding conflict with one another. Cats, it seems, had a much greater communication challenge on their plate than their competitor for our domestic affections, the dog. Dogs nudged their way into our lives much earlier than cats—although much debated, most likely somewhere between fifteen thousand and twenty-five thousand years ago. They were already well equipped with social skills and a fully developed signaling repertoire, inherited from their pack-living wolf ancestor. Compared with wolves, wildcats have relatively inexpressive poker faces, and their preferred olfactory method of communication would have been too slow-acting to work in their new face-to-face situations. They had to develop signals that were easier and quicker to read as they encountered other cats more often. Signals that could be seen from a distance as a way to declare their intentions. The tail was a logical choice.

So how did the Tail Up signal actually evolve? Some scientists have suggested that it evolved as a variation of the behavior that female cats use when receptive to mating. Known as lordosis, this rather unsubtle invitation to a male cat involves the female crouching on her front paws and holding her rump up prominently in the air with her tail slightly raised and to one side. This "presentation ritual" may have developed into the tail-raised greeting behavior over time. George Schaller, in his detailed work on the lions of the Serengeti, also described the similarity between the female sexual display in lions and their greeting behavior. He concluded that their greeting pattern was possibly a ritualized form of sexual behavior. Others have questioned this theory, citing the fact that only females display the sexual behavior, therefore making it an unlikely origin of a greeting behavior performed by both sexes.

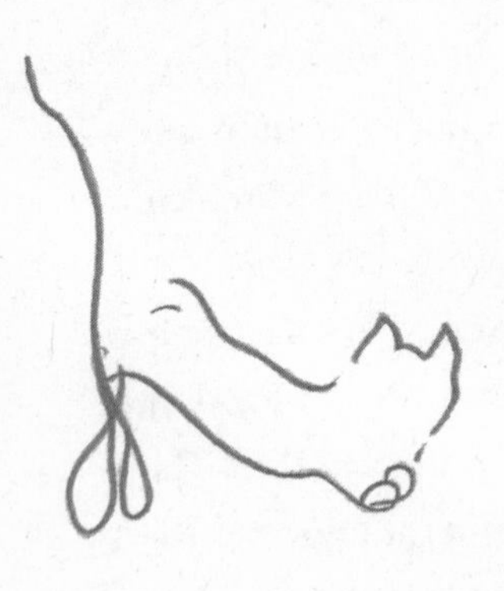

Another possibility is that the Tail Up signal may have evolved from a totally different non-signal use of the tail. Felids of many types, from leopards to the diminutive sand cat, and including the African wildcat and the domestic cat, perform a type of raised-tail behavior when they spray urine to mark their territory. This movement is similar to the Tail Up signal, although the tail is lowered as soon as the urine is expelled and, thankfully, they don't generally walk along while spraying. Nevertheless, some investigators have suggested that Tail Up could have begun this way and then developed into a signal. Snow leopards performing sprays in this way frequently go on to rub their heads on objects nearby. Could such a combination of these two behaviors have also existed in ancestral African wildcats and evolved into the tail-raised approach and head-rubbing behavior on objects, cats, and people that we see in our domestic cats today? Possibly.

However, probably the simplest and most logical explanation for the evolution of Tail Up as a signal is that cats used a behavior that worked well as kittens and kept using it into adulthood. Kittens seem to instinctively perform Tail Up when they approach their mothers. Running up, little tails held aloft like flagpoles, usually precedes friendly rubbing behavior around their mothers' chins, which in turn brings the reward of food or milk. For young African wildcats, finding themselves surrounded by adult wildcats sharing a food source, it would be a logical step to keep putting their tail up—other cats all recognize this as a peaceful sign and one of respect. Similarly, when domestic kittens, who have grown up raising their tails to their mothers, leave their mother and littermates to live as pet cats, they instinctively use the same behaviors with humans.

Retention of infantile behaviors into adulthood like this is known as neotenization, a phenomenon that occurs often in domesticated species. An example is the tendency of dogs to keep playing well into adulthood. In the cat, other behaviors carried through from kittenhood include purring and kneading with their paws (sometimes referred to as "making biscuits"). Although not social signals like Tail Up, these actions are thought to serve as a reassurance or self-comforting behavior that the cat benefits from carrying through to adulthood.

Having addressed the question of how, scientists began to ask when these hugely adaptable wildcats learned the art of tail signals. And did the wildcat have to adapt to close-up communication with people and other cats simultaneously? These lingering questions have proved hard to answer.

The change to a more social lifestyle may have happened very slowly, and wildcats may not have chosen to actively socialize with one another more than absolutely necessary. Populations of modern African wildcats, living alongside domestic cats in Saudi Arabia, often congregate around the same food sources as feral populations of domestic cats. Unlike the feral cats, however, the wildcats have not been observed to socialize in groups. Instead, they seem to merely tolerate one another. Like many animals, they may simply prefer their own company.

It may not have been until the wildcats of old were actually forced into close association, unable to avoid one another or escape, that they began to adopt new signals. Cat researchers Patrick Bateson and Dennis Turner considered when this enforced group living may have first occurred. As described in chapter one, the ancient Egyptians revered cats, worshipping cat goddesses such as Bastet and passing laws forbidding the harming or exporting of cats. Bizarrely, though, and seemingly in complete contradiction to this, these same ancient Egyptians also bred thousands and thousands of cats specifically for sacrifice to the goddesses. Evidence from vast cat cemeteries in the area shows that these unlucky cats had their necks broken at a young age, prior to being offered as tributes in temples. Bateson and Turner suggested that in these dense breeding groups where the cats lived before they were sacrificed, "the tail-up signal may have evolved rapidly to inhibit the aggression that would have been commonplace in such colonies."

Quite when cats began using Tail Up toward people is something of a sketchy picture. Evidence of cats buried next to human remains in Cyprus suggests they seem to have hung around people as early as ten thousand years ago, but whether they lived as actual pets is uncertain. Signs of true domestication by humans didn't really appear until the ancient Egyptians recorded it so artistically in wall paintings in tombs and temples around 3,500 years ago. Possibly, this era in Egypt marked the evolution of the Tail Up signal in both cat-cat and cat-human contexts.

Tops and Tails

Whenever or however Tail Up began, it is undoubtedly here to stay. Given the relatively recent (in evolutionary terms) domestication of the cat, it seems possible that, in time, the cat will evolve more visual signals in order to communicate better with other cats and people. Meanwhile, there is far more going on with a cat's tail movements than simply this friendly greeting signal. Cats are often described as aloof and uncommunicative, but they are usually saying far more than we give them credit for—and their tail movements are no exception. Among the many attempts to interpret these gestures, some authors have at times gotten a little carried away. In particular, some of the older treatises on cat behavior provide an amusing assortment of tail positions. Professor Alphonse Grimaldi, whose work was featured in a book by Marvin Clark, provided many elaborate descriptions of cat behavior and came up with these somewhat ambitious thoughts on the prophetic qualities of the cat's tail: "When it is pointed toward the fire it speaks of rain" and "When it inclines toward the door it says that its mistress may go shopping without an umbrella."

Although in reality they are unlikely to provide such detailed weather forecasts, cats, as well as performing their Tail Up signals, do actually twitch, swish, tuck, and fluff their tails all the time. Often to no avail when around humans, as people ignore or miss tail movements and persevere with their need to approach or interact with the cat regardless. No wonder so many end up with scratched hands. These tail movements generally occur alongside other visual signals, as a cat will often change their body position too. There is one particularly useful barometer of a cat's intentions at the other end of the body—the ears.

An interesting advance in the study of cat facial expressions was the development of a method of describing their facial movements according to the underlying muscles. This was inspired by a similar technique designed for humans called the Facial Action Coding System (FACS). The system has been adapted for use in a number of animal species including various primates, dogs, horses, and now cats, where it is known as CatFACS. Each "facial action" has a unique code relating to the muscles involved, some of which are similar in humans and cats. For example, "lower lip depressor" occurs in both cats and people and is caused by the same underlying muscles.

However, when it comes to ear movements, the human FACS has no descriptions listed. The musculature of our ears is not well developed and so, apart from maybe the occasional party trick of ear wiggling, we are unable to move them. Cats' ears, on the other hand, have an impressive and complex set of muscles, making them highly mobile. In fact, the CatFACS lists no fewer than seven different descriptions of ear movements for cats, which can occur in combination with one another and at different intensities: ears forward, ears adductor, ears flattener, ears rotator, ears downward, ears backward, and ears constrictor. These detailed movements were identified from lengthy, in-depth scrutiny of videos of many cats in different situations. In real time, however, such detail is very difficult for a human to differentiate. When alert and active, a cat's ears may be constantly on the move, twitching or rotating slightly to better pick up sounds from different directions. This is all compounded by the fact that the ears can rotate independently of each other.

But their ears don't just respond to sounds—cats express emotions through their ears, too, providing a useful channel of communication with others. In this respect cats are similar to horses, sheep, and dogs, all of which change their ear positions according to whether they are in negative or positive situations. Along with their varied tail movements, some knowledge of cats' different ear positions, albeit not as detailed as the ones in CatFACS, can give us clues as to how a cat is feeling.

Relaxed cats that are awake and alert will hold their ears in a neutral position, pricked upward and facing forward. This position is most likely to be observed as cats go about their daily business, wandering the house or yard. If the cat is walking, the tail may flow out level with the body or at 45 degrees to the floor. If sitting, a cat may curl the tail loosely around their body. A cat in this state might be regarded as being in a neutral mood—not focused on interacting with anyone or anything in particular.

Sometimes, as the cat sits, perhaps watching something, the

tail may begin to twitch. A twitching tail suggests slight stimulation, amusement, or excitement, a precursor sometimes to annoyance, play, or predation.

Twitching of the tail may develop into a more vigorous swishing, indicating an increased level of stimulation or annoyance, perhaps when the cat's prey or a toy has suddenly become more tantalizing. In a different context, such as being petted on a person's lap, this more agitated movement may indicate overstimulation and is the moment to stop petting to avoid shredded hands. As Alphonse Grimaldi put it, "When it lashes from side to side it signifies a war of extermination." The ears may become more energized and active here and may twitch in agitation too.

Many cats will, on occasion, experience confrontation. It may be a serious territorial dispute with another cat, or sometimes an unexpected dog or overenthusiastic person will take them by surprise. The resulting surge of adrenaline can bring about marked changes in their tail and ear positions. An angry cat, whether feeling defensive or overtly aggressive, may lift their tail into a slightly arched or upside-down U shape. The tail may be fluffed out, along with the rest of the fur, and the cat sometimes stands side-on to maximize their size and impact to their feline or other opponent. In case there is any doubt of their intentions, their ears swivel back-

ward away from the face, still raised slightly, not flattened. This overall body language suggests they are likely to attack if approached further, so a well-kept distance is the best policy. Bizarrely, kittens often perform a version of this posture when chasing and playing with a littermate, known in this playful form as "Side Step."

A frightened or submissive cat will assume a very different posture than an aggressive one. They will make themselves look smaller, sometimes crouching and tucking the tail tightly around themselves, or if standing, they will hold their tail between their legs. The ears take on a flattened position, either sticking out to the sides or, in extreme fear, flattened against the head. This ear flattening when frightened, anxious, or facing aggressive behavior from another animal seems to be a common theme in studies of other species too. In cats it is a signal for both other cats and people to be careful around them, as fear can turn to defensive aggression very quickly; cats move from one to the other of these ear, tail, and body positions with great speed as their mood changes.

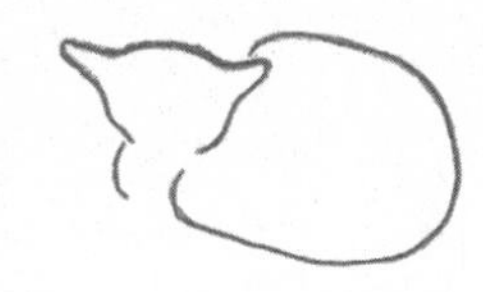

Tails of the Unexpected

My early study and others that followed from it established the importance of tail positions, especially Tail Up, in cat communication. In particular, they showed that the positions of cats' tails at the beginning of an interaction are significant in predicting the outcome. A much later study by French researcher Bertrand Deputte and colleagues looked at these beginning moments in more detail to see what roles both tail and ear positions play as one cat approaches another, and how this affects the subsequent patterns of their encounters.

The researchers studied a colony of cats living in a rescue shelter enclosure and recorded the tail and ear positions of pairs of cats as they interacted. For each cat, they recorded whether the tail was up (vertical) or down (horizontal or below) and whether the ears were neutral (described as "erect"), flattened, or positioned down and backward, much like the three different modes I've described. Results for these flattened and backward positions were later pooled in some of the analyses as "non-erect."

Their results for interactions in which both participants had their tails down (horizontal or below) are particularly interesting, given that in my earlier study some tail-down approaches had resulted in friendly interactions and some in more unfriendly ones. I hadn't recorded ear positions while my colony cats approached one another, but this new study found some interesting patterns. Regardless of the tail position of the initiating cat, if both cats had their ears erect and the recipient had their tail down, the outcome of the interaction was more likely to be positive. If, however, both cats had their tails down, and the initiator had their ears erect but the recipient didn't, then the encounter was significantly more likely to have a negative outcome. Ear positions, it seems, were important.

Turning their attention to just the ear positions, the researchers found that if both cats held their ears erect during an interaction, then it would turn out to be more friendly. If each of the interacting cats had different ear positions or if they both had non-erect ears, the interaction was likely to be unfriendly.

The study concluded that the position of

the ears was more important in cat-cat encounters than that of the tail; ear positions were a better predictor of interaction outcomes. However, it seems likely to me that cats use all the visual cues on hand to them when assessing an approaching cat. Interestingly, in every Tail Up approach made by initiating cats in the French study, the cats had erect ears, which proved to be indicators of a positive outcome. This suggests that Tail Up is rarely associated with a negative ear position in initiator cats, although it can result in non-erect ears in the recipient. Perhaps, then, it is more important for a recipient cat to read ear positions when the initiator's tail is in a more ambiguous, non-upright position.

Deputte's study also looked at interactions between these same cats and people. They found that over 97 percent of these interactions involved the cat approaching the person with Tail Up and ears erect. This was a much higher rate of Tail Up than in their cat-cat interactions, in which only 22.4 percent began in this position. One explanation for this, suggested by the authors, is that in cat-cat interactions, the position of the ears is more significant than the tail position, but with people, cats tend to use only Tail Up. This may be because humans don't understand the nuances of tail-ear combinations and so the Tail Up has become the safe, go-to behavior for cats when approaching people.

Alternatively, the different approaches may be connected to the disparity in physical size between cats and humans. People will always be larger than cats (hopefully, at least), and this alone, in any nonconflict situation, may prompt them to raise their tails and keep their ears up.

A study of how people interpret and describe different types of dog behavior showed that tail movements were the most common cues used. In general, apart from Tail Up, we seem to be less attuned to the tail positions of our cats. Interestingly, not only do dogs have a totally different language than cats, but some of their behavior patterns mean the complete opposite in the two species. Such opposite behaviors include, among others, tail wagging or swishing—a friendly or submissive signal in dogs but, depending on its intensity, a warning of predatory or aggressive behavior in cats. Given the potential for disaster when these two species meet in a tail-wagging/swishing situation, it is impressive to discover that they have apparently worked out what the other one is talking about. One study looked at four opposite behaviors performed between cats and dogs living in the same home and found that cats responded appropriately to 80 percent of such behaviors when performed by the dogs. The dogs, too, seemed to have gotten the message—they responded appropriately to 75 percent of the opposite behaviors performed by their housemate cats. The research also found that being six months old or younger when first encountering the other species improved an animal's understanding of the other's language.

This can take a while sometimes. Our young Labrador, Reggie, spent a good eight months after we first brought him home responding to our cat Bootsy's raised-tail approaches by sniffing her rear end—much to her annoyance, which she would register by whipping around and bopping him on the nose. Watching them one day, I noticed Reggie had finally worked it out—as Bootsy approached with her tail up, he sniffed her nose-to-nose as she passed by to rub all along him. In the homes involved in the aforementioned study, researchers discovered that 75 percent of cat-dog pairs displayed such nose-to-nose sniffing. We have a lot to learn from our domesticated companions.

So, Flo's Tail Up puzzle seems to be solved. It was quite simple, really. As we met that day on the hospital grounds many moons ago, Flo treated me as she would another cat. This, despite my being two-legged and (hopefully) not actually resembling a cat. As she approached, she checked me out, realized I was the older and larger of the two of us, and raised her tail to reflect that, probably hedging her bets to avoid conflict. It was a peaceful and respectful invitation to interact.

CHAPTER 5

KEEPING IN TOUCH

Too often we underestimate the power of a touch.

—Leo Buscaglia

It's a typical dinner in our household. As I eat with my family in the kitchen, our cat Bootsy sits nearby, waiting hopefully for food. Her more outgoing, confident sister, Smudge, arriving home from her daily neighborhood rounds, bursts in through the cat flap, marches up to Bootsy, and enthusiastically licks her neck. "Aaawww," we all chorus as we watch the heartwarming sisterly scene from the dinner table. Bootsy raises her paw and cuffs Smudge around the head before moving hastily away. "Bootsy!" we all exclaim indignantly. It's a set routine—the same every day. Surely this is not what is supposed to happen when cats groom each other. Where, I wonder, is the idyllic image we often see of two cats curled up together in the warm sunshine, purring and licking each other in a glorious picture of contentment?

Grooming another individual of your own species is known as allogrooming. Originally, researchers thought it was simply a

convenient arrangement used by animals to help each other clean those hard-to-reach parts of the body. Different species have developed their own techniques—birds are often seen preening each other with their beaks (allopreening), horses nibble each other, and primates tend to use their hands to pick through their grooming mate's fur. Felids, such as the domestic cat, have tongues covered with tiny backward-facing barbs called papillae, designed to enable efficient stripping of meat from the bones of their prey. When grooming, cats use their rough tongues as make-do combs, interspersed with little nibbles from their incisors to remove any stubborn dirt or parasites.

Scientists studying the behavior of various social animal species gradually began to realize that the groups they were watching were spending far more time grooming one another than could ever be needed just to keep clean. Robin Dunbar, exploring this in primates, commented, "Natural selection is an efficient process that does not often tolerate excessive amounts of slack in the biological system. Thus, the fact that such a high proportion of an animal's day can be devoted to grooming others suggests that there is a substantial benefit to be gained from doing so."

Researchers have also found that in some primate species, the larger the social group, the more allogrooming there seemed to be. In addition, allogrooming was not random within the groups they studied—certain individuals groomed certain others, but they didn't all groom everyone.

Allogrooming is now known to have a far more important function within animal societies than purely hygiene. Some animals seem to *need* to touch one another. In a diverse number of groups including cats, cows, meerkats, monkeys, ravens, voles, and vampire bats, allogrooming serves an invaluable role in maintaining social relationships. It appears to be particularly important in social systems described as "fission-fusion," in which certain members or subgroups of the population may break away from the core group for variable periods of time, only to return to the main group later. Here, there is a need to rapidly reinforce familiarity and group membership by reestablishing bonds with important friends and associates.

Allogrooming in Cats

For group-living feral cats, or indeed any two cats meeting face-to-face, close encounters are a risky business. As consummate predators, cats are armed with one of the most lethal sets of killing apparatuses (teeth, jaws, and claws) in the animal kingdom. While domestication has led to much more flexibility in the cat's social organization, this killing apparatus remains unchanged. The potential for serious injury if one cat misunderstands the other's intentions is very real. Perhaps to avoid this, cats use the Tail Up posture discussed in chapter four as their icebreaker for a conversation, enabling them to approach another cat with some confidence, waving their peace flag. But then what?

Cats, descended from solitary wildcats, have a fairly limited set of social behavior patterns from which to draw once they have approached another cat. Contrast this with dogs, who have inherited expressive faces and an elaborate range of interactive behaviors from their highly social wolf ancestors. The comparatively impassive cat, however, has had to work out ways of interacting with other cats without antagonizing them.

Zoologists David Macdonald and Peter Apps were among the earliest scientists to explore how group-living cats interact on a daily basis. Their study group was a small unneutered cat colony that lived on a rural farm. The cats lived off food provided by the farmer, supplemented with prey from their own hunting efforts. As they studied the comings and goings and daily interactions of these cats, Macdonald and Apps discovered that the activity of the group was far more complex than just a gathering of cats around a food resource. Their behavior had a social basis to it. Also, while the cats were perfectly capable of aggressive behavior toward any new, intruding individuals, aggression within the group itself was rare.

One of the most common behaviors recorded by Macdonald and Apps between cats on their farm was allogrooming. Among the adult cats, some pairs allogroomed more than others, but where it occurred it was often a reciprocal arrangement, with both cats grooming each other. The most frequently observed occurrence of grooming, however, was by mother cats toward their kittens as they licked them clean. Nestled among the hay bales, the female cats pooled their litters and washed and nursed them communally. In this warm, milky haze, the kittens first experienced being groomed by mother cats.

Mother-kitten grooming starts as a one-way process. Gradually, as kittens get older, they start to reciprocate, grooming their mothers back, along with their siblings, thus learning the art of allogrooming. This happens in the domestic setting, too, the difference being that kittens are usually separated from their mother at around two months old to begin life in a new home. Some are homed singly, others along with a littermate, and many join a household where there are already one or more resident pet cats. Sometimes two littermates adopted into a new home together will maintain their kittenhood bonds and continue to curl up together as they did with their mother, allogrooming contentedly. Or the resident cat may accept a new addition or two, sharing resting places and allogrooming. This is a classic image of cats, peacefully lying together and enjoying the mutual grooming experience. One cat being groomed by another can sometimes become so relaxed that they actually doze off.

It's magical, but sadly a scenario never displayed by my cats Bootsy and Smudge, and indeed not by many other pairs of house cats when observed by their owners.

As with many siblings or strangers thrown together, tensions may develop within the household, especially if there is competition for resources. Bonds may break down or, in the case of newly introduced cats, never develop. Nevertheless, some cats, just like Smudge, will still attempt to engage in allogrooming, even though the cats aren't the greatest of friends. To a human observer it may look like an olive branch is being offered by the groomer to the groomee, and therefore seems baffling when it is rejected by the apparently ungrateful recipient—Bootsy in my scenario.

Although there are few specific studies of allogrooming in cats, one carried out on a large indoor colony of neutered cats revealed that perhaps there is more to these exchanges than meets the eye. Surprisingly, given the apparently peaceful-looking nature of allogrooming, 35 percent of grooming interactions also contained aggressive behavior. What's more, it was mainly the cats who initiated the grooming that doled out aggressive behavior, most commonly following grooming of the partner. The study concluded that allogrooming may be a way of defusing tension between cats and, while not always successful, avoiding overt aggression. In our case at home, if the initiator of the grooming (Smudge) has exhibited such aggression toward the recipient (Bootsy) in the past, it may help explain why Bootsy and other cats living with more confident feline housemates tend to react with a knee-jerk "cuff and retreat." For Bootsy, remembering previous outcomes and anticipating Smudge's next move, it may feel safer to skip the grooming and end the encounter quickly with a swift right hook.

The association between allogrooming and aggression has been described elsewhere in the animal kingdom too. Allopreening in some species of birds, for example, may be a method of inhibiting aggressive drives. In most primates, allogrooming seems to be an affiliative, social-bonding behavior, beneficial to both parties. A study of the small nocturnal primate Garnett's bushbaby, however, showed allogrooming to be associated more with aggressive rather than friendly behaviors. The function of allogrooming, in some species, at least, may vary according to context. This might apply in the case of cats—developing between cats that already get along with one another to maintain friendly bonds (as in Macdonald and Apps's colony), serving as a functional nurturing behavior in the mother-kitten context, and being used as an attempt to defuse aggression in confined or less bonded adults (like Smudge and Bootsy).

Allorubbing

Cats and many other mammalian species living in social situations also use another form of bodily contact to communicate with each other, known as allorubbing. This is a special kind of social touch between two individuals in which one rubs a part, or various parts, of their body against the recipient. Scientists have found it hard to tell in some cases whether the purpose is the tactile experience itself or instead some kind of scent transfer from glands in the skin. In dolphins, the behavior of flipper rubbing is well studied. One dolphin swims alongside another and gently rubs its flipper over the other's. Shown by mothers toward their calves, it also occurs between adult dolphins, where it is thought to be an affiliative behavior, useful to scientists as a quantitative measure of social relationships. For dolphins, at least, given the aquatic environment of their rubbing interaction, it seems likely that the benefits are purely tactile rather than containing any scent component. Asian elephants, engaging in friendly interactions, form their trunks into a U shape as they use them to touch each other. Bonobos, closely related to chimpanzees, have a more unusual form of rubbing, reversing up to each other for a rump-to-rump rub. Cats, too, display a type of allorubbing behavior during which they rub on other cats, on objects, and, if well socialized, on people, using their heads, sides of their bodies (flanks), and sometimes their tails.

Rubbing behavior develops in kittens from around the age of four weeks, when they begin to venture out of the cozy nest they have lived in so far with their mother and littermates. Kittens will rub their heads on their mother in greeting when either party returns to the nest, raising their heads to try to reach hers as they rub. Adult cats rub in a similar way on other cats, on people, and on objects.

Although they found that allogrooming was an important behavior in their farm colony, Macdonald and Apps discovered that allorubbing between cats actually seemed to be a more significant behavior in terms of maintaining the social dynamics of the group. With this in mind, I kept a close eye out for it when I started watching my hospital and farm cats. I didn't have to wait long.

I remember the first time I saw two of my farm cats rub on each other. Penny approached Dusty face-on from the front, her tail raised invitingly. Dusty raised his tail, too, and as they reached each other, they both did a little head tilt and rubbed the sides of their faces against each other.

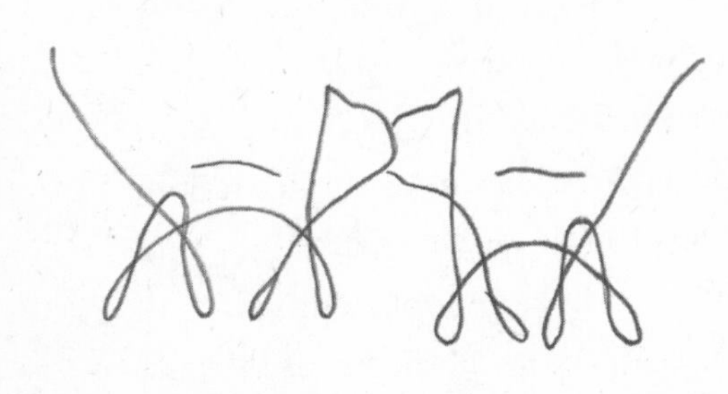

They then continued walking past each other, in opposite directions, but maintained body contact so the sides of their bodies rubbed alongside each other too. As they passed each other, running out of body to rub, their tails wrapped gently around each other, hanging on to the tactile moment as long as possible. Not content with that, they turned around and repeated the whole thing in the opposite direction. I felt a little like I had intruded on a special intimate moment between the two cats.

Unsure if this was the form that rubbing would always take, I divided my recorded observation of Penny and Dusty's little "rub fest" into segments: "rub head" (in their case, it meant the side of the face, but occasionally one cat would approach and rub their forehead on another); "rub flank" (the side of the body); and, finally, "rub tail" (wrapping the tail around that of the recipient). This proved useful—as time went by

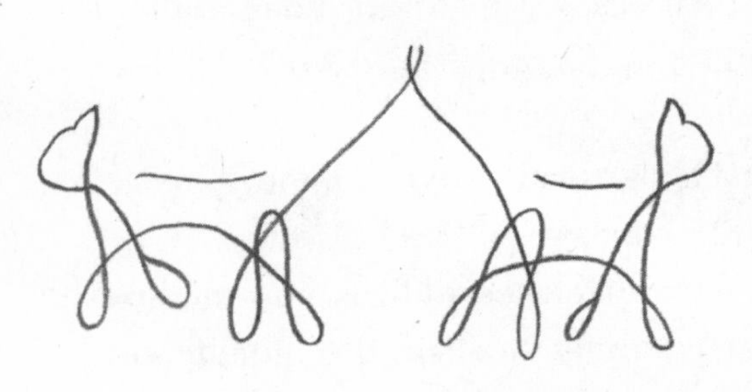

the cats taught me that rubbing was very much open to artistic license.

Sometimes mutually rubbing pairs mixed things up a bit and approached, tails raised, starting at a sort of 90-degree angle, facing the same direction and walking toward each other in a V shape till their heads met and they rubbed their faces. They would sometimes then walk along side by side, leaning on each other, tails entwined.

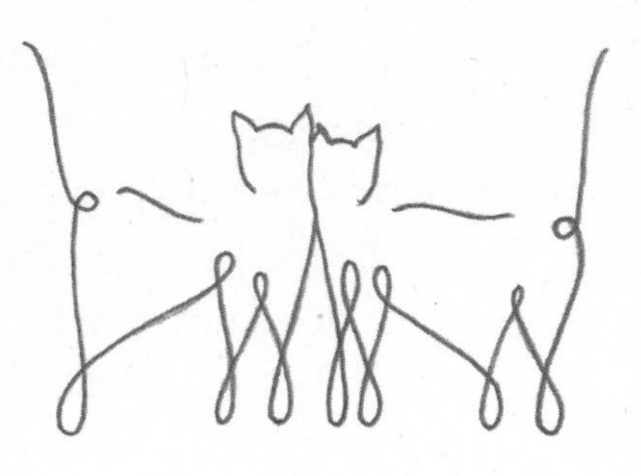

Other times rubbing was much more one-sided—one cat would approach and perform a simple head rub on the recipient cat, who would accept the rub but fail to rub back. The initiator often persisted, rubbing around the body of the other, receiving no response.

I discovered that the raised-tail exchange was a vital component of these rubbing interactions. The initiator would always approach with their tail up, but the tail response of the recipient cat dictated what followed. If the recipient raised their tail, too, the cats usually rubbed on each other simultaneously. If there was no tail-raised reply and the initiator still decided to move in for a rub, the recipient either ignored them or rubbed back only *after* this initial rub.

In Macdonald and Apps's little farm group, unlike with allogrooming, there was a definite asymmetry in the flow of rubs between cats. Kittens rubbed more on adults, adult females initiated rubs more on adult males, and some females rubbed more on other females than vice versa. In particular, there were marked differences in the rates at which the kittens rubbed on different females. Their choices were not based on which female was their mother, however, but rather according to which female nursed them more. The deal seemed to be that a female received about one rub from a kitten for three bouts of nursing. Quite how this relationship developed remains elusive—did the females nurse more in response to increased rubbing, or was the rubbing a response from the kitten to repay a generous nursing session?

I, too, found in both my hospital and farm colonies that patterns of rubbing were uneven, with some cats initiating more and some cats receiving more rubs. My groups of cats were all neutered and so there was no kitten-mother element, making it hard to compare directly to Macdonald and Apps's findings, but there were definitely preferred rubbing partners among the adults.

As with Macdonald and Apps's colony, rubbing in my colonies seemed to flow mainly between cats of unequal size or apparent status, from the smaller or weaker to the larger or stronger ones (see opposite page). This is much like Tail Up in chapter four—perhaps not surprising, given the close association of the two behaviors. For example, Frank, from the hospital group of five cats, was the one who was present for the least amount of time during my observations. And yet he was the recipient of almost half of the total number of head rubs performed by members of the group. Nell, on the other hand, who was there most of the time, rarely received a rub from other cats. A large, imposing cat, Frank patrolled a much wider area of the hospital grounds than the small courtyard where I watched. By his demeanor alone, he commanded a higher status than the smaller, petite female Nell. Betty made no secret of her preference for rubbing on Frank—or, when he wasn't around, Tabitha.

I made an interesting new discovery while analyzing my hospital-cat and farm-cat observations: When pairs of cats sat

Rates of head rubbing between cats in the hospital colony—adjusted for time each pair was present together during observation periods

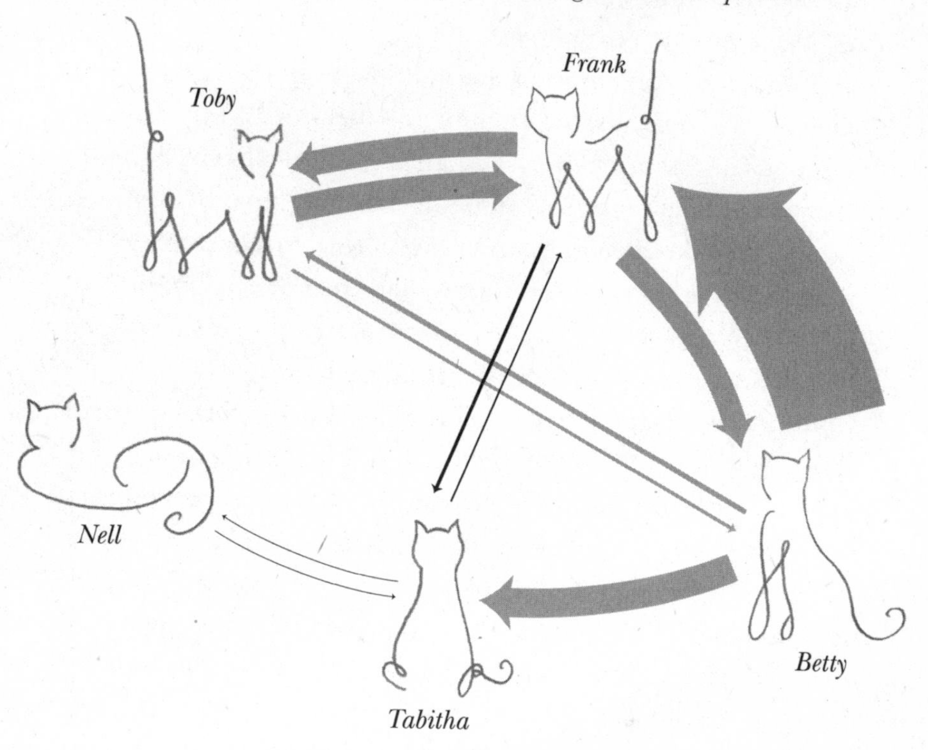

close together and occasionally allogroomed, they were unlikely to rub directly before or after this behavior. This suggests that allogrooming and allorubbing might be different types of social bonding behaviors for group-living cats. Allogrooming may be a way of maintaining close proximity between cats by minimizing tension. Allorubbing, on the other hand, may be used more during greetings, when cats haven't seen one another for some time, such as in a feral colony in order to reestablish social bonds—a way of welcoming them back into the social group. So those that hang around the core area of the group, often the females, may not need to rub on each other so frequently. Colony members—such as the larger males—who spend less time with the main group receive more rubbing.

In domestic situations where multiple cats live together with one or more humans, the amount and focus of allorubbing may be quite different from that in an outdoor colony. There are many anecdotal reports of domestic cats rubbing more on their owners than on each other. During my own experimental studies of rubbing in indoor cats, I found that when more than one cat was present along with a familiar person, the cats rubbed on the person and on an object, but never on each other.

Researchers Kimberly Barry and Sharon Crowell-Davis visited sixty households that contained pairs of neutered indoor-only cats and studied their interactions. Allorubbing occurred very infrequently in the male-male and male-female pairs and not at all in the twenty female-only pairs over the ten hours each pair was observed. The authors suggest this may reflect how such cats are stable group members who don't move far out of each other's vicinity (being indoors all the time), in contrast to feral cats, who come and go from the core colony area. Outdoor pet cats or those who go in and out may have more of a need to greet and rub on their return to the home.

Scientists have had much discussion about what is actually going on when cats rub, with some suggesting that there may be different kinds of rubs according to the situation. Sometimes rubbing may be aimed at leaving a scent mark on an object for other cats to find at their leisure. At other times, it is used as a more direct social behavior, involving tactile contact between two cats or acting as a visual display as one cat rubs on an object while another cat watches.

As described in chapter two, cats are well endowed with scent glands in their skin, concentrated in certain areas of their body.

The glands are particularly numerous on the chin and along the side of the face, with some more at the base of the tail and even between the pads of cats' paws. Many felid species, from snow leopards (*Panthera uncia*) to smaller ones such as the fishing cat (*Prionailurus viverrinus*), have a habit of rubbing the sides of their faces against prominent objects such as trees and rocks in their environment. A domestic cat's "scent post" might be a branch or fence post outdoors or, when indoors, the edge of a cupboard, doorway, or box. Often preceded by the cat sniffing the object, such deliberate rubbing or marking by both wild and domestic cats is most likely designed to leave a scent behind. Domestic cats also rub nearby objects in social situations, such as after an aggressive encounter with another cat. On these occasions the rubbing seems to be more of a ritualized visual display, although there may also be a scent-depositing element to it.

Some say that cat-to-cat rubbing is also carried out in order to transfer scents. One cat rubbing on another, however, involves far more elements than a cat rubbing on an inanimate object. For a start, there are two cats and so two separate sources of scent. If scent marking is the sole purpose of such rubbing, is only one of the cats leaving a scent mark on the other, and if so, which is depositing and which is collecting the scent? Or are they actually deliberately mixing their scents to produce some kind of joint or group odor? The latter has been proposed for badgers, for example, following studies of their scent marking. With respect to cats the answers to these questions remain elusive. Interestingly, though, unlike when they approach an object, cats rarely stop to sniff each other's heads, or indeed any other part of the body, before allorubbing. This raises the possibility that perhaps when cats allorub, first with the head, followed by the flank and the tail, there is another, non-scent-based element to the interaction—a tactile one. Maybe it simply feels good to touch each other.

The Magic of Touch

Over the years, in their search to unveil the mysteries of touch, scientists have examined the skin of mammals in more and more detail. Hugely underrated in importance, the skin is actually the largest organ of the mammalian body. It is home to a whole host of sensory nerve receptors, each responsible for responding to a different sensation. These include, among others, thermoreceptors, which sense temperature; pruriceptors, which register an itch; and nociceptors, which respond to painful stimuli. There are also at least seven different types of so-called low-threshold mechanoreceptors (LTMRs), which respond to touch, supplying the brain with information on shape, texture, pressure, and other tactile features. Each of the neurons transmitting this information is surrounded by a myelin sheath, an insulating feature that enables messages to be conducted to the sensory cortex of the brain at rapid speed—an essential feature to enable fast reactions to potentially dangerous stimuli.

Cats are particularly tuned in to touch, blessed with an extra connection to it via their famous whiskers. Most cat owners will, at some stage, discover somewhere around their house a long, thick, tapered whisker shed by their cat. It will most likely have come from the clumps that protrude from the two sides of a cat's upper lip, slightly reminiscent of a long, wispy mustache, giving them their name, the mystacial whiskers. These are the largest and most familiar set of what are technically known as vibrissae, located in tufts at various locations on the cat's body. As well as the mystacials, there is a clump over each eye, two small tufts on each cheek, and clumps on the backs of the front legs.

The whiskers are an important source of tactile information for cats, helping them navigate and hunt efficiently. Although they contain no sensory component themselves, the roots of the whiskers are buried deep in the skin, where they are surrounded by sensory receptors. These receptors send information to the brain regarding the position and movements of the whiskers as they brush against objects. Scientists believe the receptors are also sensitive to air currents, making them particularly useful in the dark.

Well endowed with muscles around their roots, the whiskers are surprisingly mobile. This is particularly noticeable in the mystacials, which move as a group rather than individually, with the left and right sets able to move independently of each other. This movement may serve a practical purpose—for example, when hunting, as cats get close to their target and their natural long-sightedness makes focusing difficult, the mystacials sweep forward to provide tactile information on the location of their prey. Cats may also perform this forward whiskery sweep in response to something else that seems interesting—say, a human hand held out for them to sniff. The facial action coding system developed for cats (CatFACS) describes this forward motion as "whiskers protractor." It also identifies a movement known as "whiskers retractor," in which the whiskers flatten back against the face, and "whiskers raiser," in which they are directed upward. The detailed dynamics of these movements are hard for humans to follow visually, even when we look closely at the cat's face. But we do know that typically, cats will hold their whiskers out to the side in a neutral position when relaxed, while stressed cats will pull them backward and flatten them more. Much like with ears, reading the whole body language is a useful overall way of gauging a cat's mood.

In 1939, while investigating skin receptors, neurophysiologist Yngve Zotterman discovered, nestling among the many rapid-firing myelinated nerve fibers, a different sort of receptor, subsequently given the tongue-tying name C-fiber low-threshold mechanoreceptors (CLTMs). He found that these, unusually, had no myelin sheaths, and so responses from them traveled far more slowly. Interestingly, Zotterman was working with none other than the furry skin of cats when he made his discovery. It was not until fifty years later, using new neurological techniques, that an equivalent structure was identified in the hairy areas of human skin. These human versions were christened more simply "C-tactile (CT) afferents."

Much research has followed on the nature of these CT afferent nerve fibers, particularly their role in social touch. They are most effectively stimulated by slow, gentle stroking of the skin, as a mother might caress her baby, one friend might comfort another, or a person might pet an animal. Studies have shown that stroking at a low pressure, at body temperature, and at a particular speed (between one and ten centimeters per second) creates the maximum response from these CT afferent fibers. People seem to instinctively stroke at these most effective speeds when caressing their partners or babies. And their pets.

Social touch evidently has a calming and bonding effect on many groups of animals—but how does it work? Unlike the myelinated receptors that enable the brain to discriminate different types of touch and respond accordingly, imaging research has shown that the CT afferents stimulate a different area of the brain, known as the insula, a structure particularly associated with generating feelings of pleasure. Studies of stroking show that the frequency with which CT afferents fire their messages is positively correlated with people's ratings of tactile pleasantness. In other words, gentle stroking makes the recipient feel good. CT afferents have now been found in every mammalian species investigated. No wonder, then, that so many animals like to be touched, groomed, rubbed, or stroked by others.

Research began to focus on how this feel-good factor works—how a gentle lick, stroke, rub, or brush of the skin in humans and animals is such an effective hedonic stimulus. The answer was found in the biochemical side of this phenomenon. It turns out that social grooming and rubbing in animals and gentle stroking in humans are linked to the release of a whole suite of neurochemicals that affect emotions. The one that has received the most attention is oxytocin, also known as the love hormone. Particularly important for the bonding of mothers and infants, it also has a significant effect on social behavior. While it was once thought to have an exclusively positive effect, increasing trust and friendly relations with social partners, studies in humans have shown that oxytocin release actually has a variable effect depending on the context. It seems to heighten an individual's awareness of social information, enabling them to react favorably to recognized allies but more defensively when they are less sure of the integrity of the interacting partner. Evidence of this has been found in nonhuman animals as well, and can result in a kind of social selectivity effect that reduces sociality toward unfamiliar individuals. For colony cats encountering a familiar feline face, a tactile rub may increase oxytocin in both cats and enhance their social bond, while at the same time making them alert to unfamiliar cats.

Alongside this increase in oxytocin, social touching has been shown in some species to reduce levels of the stress hormone cortisol, and to lower heart rate and blood pressure. Another group of hormones known as endorphins are also closely associated with the social-touch "feel-good" experience. Endorphins released during exercise or massage of the skin, for example, produce a mild natural opiate-like effect, bringing a feeling of well-being and calm.

Although these physiological effects have not been studied directly in connection with tactile interactions between cats, it seems likely that they would experience something similar. It certainly appears that the tactile experiences of allogrooming and allorubbing are vital components of the social behavior repertoire cats have developed—reducing tension and strengthening bonds between partners.

Rubbing on People

Thought by many owners to be one of their most endearing behaviors, cat-to-human rubbing, just as with cat-to-cat rubbing, has many variations. It often begins with the trusty Tail Up as the cat approaches the person.

Having announced their peaceful intentions, the cat may then rub their head, flank, and sometimes their tail along the person's legs, often wrapping the tail around the recipient's leg, as they might on another cat's tail.

Depending on how the person is standing, the cat may weave in and out of their legs, figure-eight-style, in an energetic display of rubbing.

Cats often attempt more ambitious moves, too—what we in my family call a "jumpy rub"—in which they attempt to rub their heads a little higher up the person's leg by lifting their front legs and performing a sort of hopping move. Perhaps this

is an attempt to reach what may be their preferred rubbing target—the person's head.

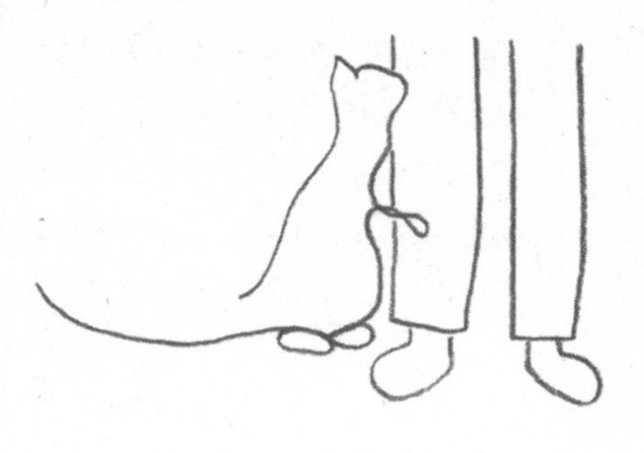

When interacting with a person, domestic cats often extend their repertoire to rub their heads on nearby objects such as cupboard corners, boxes, or basically any inanimate object that doesn't move when rubbed. This is similar to the object rubbing seen in some rubbing encounters between cats, although in cat-human encounters it seems to be particularly exaggerated.

I decided to carry out some investigations to find out more about cats' rubbing behavior around people. Studies of cats rubbing on humans were relatively rare when I first began mine. In one simple and elegant investigation of human influence on the behavior of cats, Claudia Mertens and Dennis Turner ran an experiment to see how cats reacted in staged encounters with people they didn't know. For the first five minutes, the person sat reading a book and ignored the cat. Then, for the next five minutes, the person interacted with the cat freely. The researchers found that, overall, the cat rubbed their head significantly more on the person during the interactive period. Head rubbing is therefore evidently an important component of social interactions between cats and humans—for the cats, at least.

To explore this further, I carried out a study similar to that of Claudia Mertens but with a few variations. In contrast to Mertens's experiments, the person participating in my tests was familiar to the cats, and I also included a wooden box in the room for the cat to rub on if they wanted to. As well as seeing how cats altered their rates of rubbing on the person when they received attention from them, I also wanted to investigate how this affected their rubbing on a nearby object. During each video-recorded test, a cat was introduced to the room containing the person and the box. The person then either ignored them or interacted in a consistent way (stroking for twenty seconds every minute and talking freely) for a period of five minutes.

Analysis of the videos showed some interesting results. The amount of object rubbing by the cats increased significantly from an average of 9 object rubs while the person ignored them to 23.8 rubs when the person interacted with them. The cats rubbed on the person, too, but less so than on the box. They averaged 5.3 rubs on the person when no contact was given. Interestingly, though, when the person interacted with them, this human-directed rubbing only increased to an average of 6.9 rubs—a nonsignificant change. So, while interaction from the person led to many more rubs on the object, it only slightly increased rubs on the person. Mertens's study did not record object rubbing by the cats, so it is difficult to compare the studies directly. Possibly cats rub more in general when a person gives them attention and, depending on what is available for rubbing, direct their rubs to different targets.

Why do cats rub so much on objects when interacting with people? While a definitive answer will probably always remain a well-kept cat secret, looking at the interaction from the cat's point of view gives some clues. Rubbing on a human's legs is quite different compared with rubbing on a cat of similar size, height, and shape. Up close to a person's legs and positioned far below them on the ground, the cat can't properly see the recipient's face or reaction while rubbing on them. It seems logical to move away slightly and rub on something else nearby while observing the person's face and body language. This enables them to see whether their greeting has been noticed, maybe make eye contact, or even throw in a meow to request whatever it is they want. A visual display, perhaps, designed to keep the greeting going while checking whether they have the person's attention. This redirected rubbing is such a subtle yet clever adaptation of cats' behavior, we barely notice it. Yet every friendly domestic cat I have ever met will include it as a component of their rubbing interaction.

The perils of overlooking such elegant behavioral nuances were revealed in a paper enticingly titled "Tripping over the Cat" by Bruce Moore and Susan Stuttard. The authors revisited some classic and much-acclaimed experiments conducted by scientists Edwin Guthrie and George Horton back in 1946, which had purported to illustrate various types of learning in cats. The cats had been placed inside a puzzle box, from which the means of escape was to maneuver a vertical rod. Cats were remarkably good at this and showed a stereotyped pattern of movements over and over again—apparently, according to Guthrie and Horton, "learning" how to move the rod and escape. What Guthrie and Horton failed to realize was that each time they placed a cat in the box, they also had people sitting nearby, unconcealed, observing the experiment. What resulted was the cats' natural instinct to perform a greeting rubbing display toward the observers. Unable to rub on the observers themselves, they directed their rubs to the nearest object, the escape rod, and—hey, presto—became masters of the escape! To illustrate that this had been the true explanation, Moore and Stuttard re-created the experiments of Guthrie and Horton, but recorded reactions of the cat with and without a human observer present. When watched out of sight, the cats did not perform the rod-rubbing routine. Guthrie's cats had therefore not shown learning at all—they were just performing their natural cat-human behavior.

There are other subtle differences in rubbing behavior that owners may be unaware of. Pet cats that are allowed outdoors, for example, rub on people more than indoor-only cats do. This is probably a reflection of their need, as in cat colonies, to reinforce social bonds on reunion with animals or people to whom they are socialized (in this case, their owners). When owners leave the house, cats rub on them on their return, but, interestingly, the amount of rubbing does not increase after longer periods of separation. This suggests that there may be an element of ritualization in the cat-owner rubbing greeting display rather than it being proportional to how much the cat has missed the owner. This contrasts with the greeting of owners by dogs after different periods of absence. Dogs have been shown to display much more intense greeting behavior (tail wagging and interaction with their owner) after the owner is away for a longer period of time.

Cats don't restrict their rubbing behavior to humans and other cats. Many instances of cat-dog and even cat-horse rubbing have been described. Our cat Bootsy, for example, despite her rocky relationship with her sister, Smudge, had a particularly strong bond with our golden retriever, Alfie. She would enthusiastically rub on him whenever they met in the house after a period apart. Smudge, on the other hand, would give Alfie a wide berth or alternatively a quick swipe across the nose—she never rubbed on him and most definitely did not consider him to be part of her social group.

It is often suggested that rubbing on their owner is an adaptation by cats to obtain investment. What investment are they looking for? Many people say they are only looking for food, and it's true they tend to rub on whoever feeds them at a high rate around dinnertime. A hark back to kittenhood, perhaps, where, as on Macdonald and Apps's farm, increased rubbing correlated with more nursing.

However, many cats also rub around their owners when they are not hungry, especially as a form of greeting behavior when the owner or the cat returns to the home. During a set of tests designed to assess their attachment to their owners, Claudia Edwards and co-researchers recorded how much head rubbing cats displayed when alone, with their owner, and with a stranger. They found that head rubbing (on the person and objects combined) increased significantly when with a stranger compared with when they were alone. The cats also rubbed significantly more when with their owner than when they were with the stranger.

Stroking a Cat

The typical response by a person to a cat rubbing around their legs is to bend down and stroke or pet them. Petting a cat in this way is probably the closest behavior humans have to providing a reciprocal rub for a cat, and sometimes it may be that this is all the cat is hoping for. These leg-rubbing and stroking interactions are most likely to be asymmetric in nature, as it seems unlikely that a person would bend and stroke a cat at the same rate as the cat rubs—it would be way too awkward. The balance is likely to be reversed when a cat is sitting on a person's lap, where the human can stroke more easily than the cat can rub. This may help explain why cats seem to need to build a particular level of trust before they will curl up on a lap—in so doing they inevitably lose their control of the tactile interaction, something many cats dislike.

One of my favorite words used in the study of human-animal

interactions is "gentling." It refers to a combination of stroking/patting and calmly resting a hand on an animal, with or without speaking quietly. It's a technique long used to enhance bonding between humans and many animal species, including farm, laboratory, and companion animals. Looking at the combinations of talking and touching in human-pet interactions, researchers found that the sound of their owner's voice on returning from being separated caused a rush of feel-good oxytocin in dogs. Interestingly, this effect was much longer lasting if the owner combined talking with stroking—the tactile part was important. Watching dogs roll over to have their soft underbelly rubbed, there seems little doubt that they enjoy such contact.

With cats, the effects of a combination of talking and stroking may vary depending on circumstances. A study by Nadine Gourkow and colleagues looked at the physiological effects of gentling (both stroking and gentle vocalization) on particularly stressed cats housed in animal shelters. They found that the cats that received this treatment showed increased levels of secretory immunoglobulin A, and subsequently exhibited better resistance to upper respiratory infections. The non-gentled group of cats turned out to be over twice as likely to develop these types of infections over time compared with the gentled group. However, another group of researchers, looking at the best combination of stroking and talking to soothe cats in shelters, found that, for some cats at least, the stroking was more effective alone, without added talking. This may have been specific to the study; possibly the cats were too unfamiliar with the people who did the talking, or they may not have found their voices soothing, or simply the cats may have felt that the people were being too pushy.

In a home environment, pet cats frequently solicit tactile interaction by rubbing on their owner, suggesting that for them, being petted is an important part of conversation, in addition to any verbal-exchange ritual they may have with their owner. In the same way that cats and their owners develop their own vocal conversations, many also have ritualized routines of behavior that include rubbing and stroking. Owners often describe how their cat "invites" them to start this by rubbing on a leg or a nearby object and looking up at them, or by jumping onto their lap. Some cats will even lead their owner to specific areas of the home where they prefer to start these rubbing/stroking interactions, presumably because they associate certain places with that activity.

Cats have areas of their bodies where they prefer people to stroke them. Sarah Ellis and co-researchers looked at this in more detail. Choosing eight areas of the body listed by owners as regions where they petted their cat, they tested how cats reacted when stroked in each, recording both negative (aggressive or avoidant) and positive (friendly) behaviors. Touching the tail region resulted in much higher incidences of negative behaviors by the cats compared with petting areas around the cheeks and chin (perioral area) and between the eyes and ears (temporal region). In other words, it is best to stick to the head regions and avoid the tail when petting cats. The stomach is one of those mysterious areas when it comes to stroking—many owners describe how their cat will roll ecstatically around on the floor when being petted, apparently inviting them to stroke their soft underbelly, only to clamp on to the offending hand if it attempts such an intrusion.

Some cats will guide the person stroking them to the specific areas of their body they want to be petted, maneuvering their head or body to maximum effect. Interestingly, cats guide or at least prefer people to pet them in the same areas that they would be head-rubbed on or allogroomed by another cat. This is true of many animals—some other species also most enjoy human contact in the body areas that are normally groomed or rubbed on by conspecifics (animals of the same species). Dairy cows, for example, have a preference for being massaged by a human along their withers (the ridge between the shoulder bones), the same area where cows choose to gently nibble and groom each other.

Why do cats want us to stroke them? The areas of the head where they choose to be stroked are endowed with scent glands, and some scientists suggest that they are trying to imbue us with as much of their scent as possible by encouraging us to rub our hands over them. But there are glands at the tail as well, and most cats really don't seem to want us near those. Also, as with other cats, whose scent is landing on whom when our hands rub over their faces? Perhaps the aim is to mix our scents. If it's only about scent marking, then they could just rub on our legs as they pass by rather than waiting around for the human to bestow them with a reciprocal stroke. Cats' tendency to continue rubbing, backward and forward, when we respond to their initial advances suggests that they want this social interaction to go both ways. They seem to genuinely enjoy it.

What's in it for us? A study that looked into stroking between human partners discovered that being stroked was both more pleasant and more intense compared with being the one doing the stroking. Nevertheless, stroking was still a pleasant experience for the stroker. This is certainly the case when it comes to humans stroking cats. Researchers looking at women interacting with their cats discovered that some specific interactive behaviors correlated with a rise in oxytocin levels in the saliva of the women, while others did not. Gentle petting, hugging, kissing the cat, and initiation of contact by the cat, for example, were all correlated with higher resulting oxytocin levels, whereas purring from the cat, or the woman talking in a gentle baby voice, did not produce the same effect. It appeared that the tactile behaviors were the ones that produced the feel-good effect in the women. A similar effect was found in another study that showed that stroking the warm fur of a cat produced a positive mood in women. This was reflected in the activation of the part of the brain called the inferior frontal gyrus, an area that is known to be associated with emotional and social communication. These last two studies focused specifically on women, exploring the maternal aspects of cat ownership. The ways in which women interact with cats are explored further in chapter eight.

While many cat-human rubbing interactions begin with the cat weaving around the owner's leg, others begin with the cat jumping onto the owner's lap and inviting the person to pet them. Here the cat may rest, apparently enjoying the stroking, perhaps purring luxuriously, body relaxed, eyes sleepily closed. Sometimes there is a sudden subtle change in body language—the cat is not so relaxed, the muscles tense, the ears swivel slightly back, and the tail starts to swish. The owner, idly continuing the stroking and lulled into a false sense of security, may be distracted by the TV or a book or whatever they were doing when the cat first landed on them. They may not notice until, seemingly out of the blue, the cat lashes out and claws or bites them before leaping off and retiring to a point somewhere across the room, where they haughtily groom themselves. Left sitting bewildered, the owner may wonder what on earth just happened.

"Petting and biting syndrome," as it is known, is surprisingly common in cats. With individuals that are prone to reacting like this, it is important to look out for the warning signs. Many cats simply do not like to be petted, even if properly socialized to humans. In a survey of cat owners from Brazil, researchers asked about cats' social behavior in the home. Owners' responses indicated that 87 percent of cats apparently liked to be stroked. And yet, when asked about aggressive behaviors in their cat, 21 percent of owners reported aggression in their cat when petted or put onto a lap. This was, in fact, the most common context in which aggression occurred. Something doesn't quite add up—why do some cats that usually enjoy being petted suddenly get fed up with it and lash out? If petting by a human is, for cats, the equivalent of being groomed by another cat, perhaps the petter is not doing it correctly, possibly stroking them in those areas not normally groomed by another cat, such as the tail region.

Another possibility, supported by research on stroking in humans, is that the petting goes on for too long and reaches a point where the cat really isn't enjoying it anymore. The CT afferent fibers that relay the stroking sensation to the brain have been shown to tire and reduce their rate of firing with repeated stimulation. In humans, this is matched by a gradual decrease in both feelings of pleasantness and the desire for stroking to continue. Assuming a similar process occurs in cats, such "stroking satiety" might account for the change in behavior of some cats when the balance between wanting to be petted and enjoying it tips over into a need to escape.

Allorubbing and its human equivalent of stroking is something of a success story for the cat-human relationship. Both have adapted their social touch behaviors to enable a mutually rewarding interspecies interaction, although cats have most definitely worked harder on it. Rubbing by cats has become one of the favorite feline behaviors of owners and nonowners alike, hard to miss as they use it to literally nudge us into a response. One interesting study looked at facial movements and other behavioral signals given by cats in adoption centers and how people reacted to them. They discovered that rubbing was the only behavior that affected the cats' adoption rate—cats that showed higher frequencies of rubbing were adopted more quickly.

Is rubbing by cats on people simply friendly behavior or something more calculated? Ask pretty much any cat owner, and they will say the former. As the famous British vet James Herriot once said: "I have felt cats rubbing their faces against mine and touching my cheek with claws carefully sheathed. These things, to me, are expressions of love."

I like to think he was right.

CHAPTER 6

SEEING EYE TO EYE

> *An animal's eyes have the power to speak a great language.*
>
> —Martin Buber

As I write this, I am half watching our cat Bootsy outside in the garden. She's quite distracting. It's October, very blustery, and leaves are blowing everywhere. Bootsy is in heaven. Every time a leaf flutters up into the air, her eyes dart in pursuit and she leaps and pounces on it. To say this is her favorite hobby might be an exaggeration, but "chasing things that move" is certainly right up there on the list, along with lying in the sun, washing herself, and, of course, watching birds out the window.

The lives of pampered domestic cats like Bootsy are very far removed from their distant ancestors. Yet their senses and instincts remain much the same. Of all the senses, vision is one of the last to mature in kittens, with olfaction and touch responsible for guiding

their earliest activities. That's not to say vision isn't important to cats; it's just important in a whole different way than it is for humans. Both cats and humans use their eyes to see what is going on around them, but our different evolutionary backgrounds mean we are not necessarily looking for the same things. Human vision is primed for daytime use: seeing everything in glorious color, watching what other people are doing, and using our eyes as a means of communication. The cat's visual system, on the other hand, is superbly designed for what was the single most important objective for their wildcat ancestors—catching the next meal. Unlike Bootsy's autumn leaves, the rodents on which cats naturally prey are mostly active at dawn and dusk, so this is when cats—known as crepuscular hunters—traditionally tend to hunt. Cats need to be able to see and catch fast-moving prey in the near-dark at the beginning and end of the day.

Although cats possess the same basic mammalian eye structure as humans, their eyes also have a number of significant adaptations that improve vision in these darker settings, maximizing the amount of light entering them to help distinguish images in semidarkness. For a start, they have a whole extra layer at the back of their eyes, which reflects any incoming light back to the retina for a second chance to stimulate the photoreceptor cells located there. Known as the tapetum lucidum, this layer is what gives cats' eyes their "glow-in-the-dark" effect.

The photoreceptor cells in the retina are the same two types that we have, but cats have them in different proportions than we do. One type, the cones, are active in bright light and detect color. Humans have three kinds of cones, each sensitive to either blue, green, or red hues. Cats have relatively few cones and none of the red-sensitive variety—they see mostly blue and yellowy-green colors. Their resulting view of the world is therefore much more muted color-wise compared with humans. The other type of photoreceptors are called rods. These enable black-and-white vision only, but they function in low-light conditions, ideal for seeing in near-darkness. While we have relatively few of these rods, rendering our eyesight fairly useless as soon as the sun begins to set, cats have many of them. In addition, the pupil of the cat can open much wider than ours, allowing more light into the eye as darkness falls.

Cats can't see in complete darkness, but light from the moon—or, for the modern-day cat, streetlamps—provides enough illumination to see what they need. In bright light, their pupils rapidly shrink to narrow vertical slits, an adaptation that protects the eyes while maintaining the ability to focus on any potential prey under daylight conditions. Pupillary changes can also occur for reasons other than adjustment to the light. Emotional rushes of fear or excitement result in a wide-eyed look with dilated pupils, while a calm cat generally displays more narrowed, slit-shaped pupils.

Cats also have a slightly larger visual field than humans, panning out to around 200 degrees compared with our 180. They pay close attention to their extra peripheral vision, tending to ignore stationary subjects but quick to chase anything that moves, as illustrated by Bootsy and her many leaves. Their eyes are able to track a traveling object (or mouse) by scanning with a series of small jerks called saccades that stop the image from looking blurred.

Reduced color vision is probably of little consequence to cats—they are more concerned with contrast or patterns than with actual colors when looking for prey. So, for a cat, a toy ball with a black-and-white zigzag pattern will be more exciting to chase than a brightly colored all-red one. Many owners wonder why their cats appear confused and find it difficult to locate a treat placed on the floor right in front of them. This is because, due to the musculature of their eyes, they struggle to focus on anything closer than about ten inches, tending to use their whiskers or sense of smell to locate things once they get up close. Their distance vision is also not great—farther than twenty feet away, the world appears a bit fuzzy to them.

Cats' eyes may be hardwired for hunting, but they still keep a close eye on everything else going on around them, a skill they develop from a very early age.

Look and Learn

Young cat! If you keep
Your eyes open enough,
Oh, the stuff you will learn!

—Dr. Seuss, *I Can Read with My Eyes Shut!*

Just as with any species, the ability to learn from what they see, known as observational learning, is essential for cats, especially young ones. Kittens that are born away from the comfort of a human home need to learn from their mothers how to hunt or they are unlikely to survive into adulthood. After initially bringing them dead prey to eat, the mother cat starts to bring live prey to the nest and will show her kittens how to kill it. Once they have watched her do this a few times, the lessons progress to letting them have a go at killing the prey themselves.

Researchers interested in observational learning in kittens have shown that they are able to watch and learn a variety of skills from their mothers as well as from other cats, even in quite unnatural circumstances, far removed from the aforementioned situation. Experiments by Phyllis Chesler that looked at whether kittens could learn to press a lever found that left to their own devices, they hardly ever worked it out. However, when they were able to observe another cat doing it, the kittens gradually learned to perform the same action. This observational effect was greater if the demonstrating cat was the kitten's mother. Once a kitten had gotten the gist of what it should do, the identity of the demonstrator cat had no effect on how quickly the kitten subsequently improved their success rate. Therefore, the important element in the learning process was how closely the kittens paid attention to the demonstrating cat in the first place, with the mother cat holding a greater pull in this respect. The watching skill seems to be retained into adulthood, too—a study of observational learning in adult cats found again that having another cat to watch sped up the learning of a task.

Cats seem to put their ability to look and learn to good use whatever their living circumstances. For group-living cats, avoiding confrontation while maintaining access to the all-important resources of food and shelter is the priority. In farm-cat colonies, for example, spotting another cat returning with prey from a successful hunting spree will give a cat a heads-up on where to hunt themselves next time. The focus of pet cats' observational learning may be slightly different from that of cats who have to forage for food and live on their wits. They learn from people instead, watching what we do and remembering which of our activities bring good results, such as when they see us reach for the can opener and a saucer of tuna soon appears.

Now You See It . . . Now You Don't

If someone takes an object and hides it behind their back, does it still exist? As adult humans we know it does, and we should look behind the person's back to find the object. As infants, though, we start off thinking that the object, once hidden, has gone. That is why peekaboo is such a fun game for babies—each time we disappear behind our hands we have apparently gone, before we reappear to renewed amazement. The theory of object permanence was proposed by a Swiss psychologist named Jean Piaget, who first looked at its development in human infants. He tested babies by removing one of their toys and, while they were watching, hiding it under a blanket. This was known as a visible displacement test. Piaget discovered that over the first two years of life, children gradually develop the understanding that although they can no longer see it, the toy still exists. They become able to hold a mental representation of the object in their mind while it is missing, and they will begin to search for the item where they last saw it. A more complicated "invisible displacement" test involves placing the toy in a container in front of the child, moving the container behind a screen, then removing and leaving the toy behind the screen. The child is then shown the now-empty container, and they must work out that the toy was taken out when the container was behind the screen, and therefore they should look for it there.

Animal-cognition scientists realized that the ability to understand object permanence would be a hugely advantageous skill for many species of animal, not just humans. Knowing that a predator or prey that recently disappeared from view is now hiding behind a particular rock can be life-changing information, determining whether, as prey, you escape with your life or, as predator, you catch the next meal.

Following Piaget's work and using similar methods to those he used with children, researchers have subsequently tested other nonhuman animal species for their perception of object permanence. Cats, when tested, have been shown to succeed in visible displacement tests—in other words, they understand that when an interesting object goes out of sight, it remains in the place it was last seen. They acquire this ability by the time they are six to seven weeks old. Researchers have also tried carrying out Piaget's invisible displacement tests on cats. This generally proved too difficult for them, though—when presented toward the end of a test with the now-empty container, cats seemed unable to deduce that the object that had been inside it had been left behind the screen. They would search near the container instead.

Although perhaps surprising to us, rather than revealing anything about the cats' intelligence, these results may simply reflect what cats need to know in the wild. The visible test scenario, representing when a prey animal hides from sight, is one that they might realistically encounter in their natural environment. In fact, cats can do several of these tests in a row with the destination hiding place changing each time. Provided they actually see the object as it is being hidden, the cats keep visiting the correct place to look for it. This may happen when a mouse runs and hides, runs again and hides, and so on. Much less likely, however, is any natural situation in which prey or a predator becomes invisibly displaced.

An important factor for a predator waiting for hidden prey to reappear, and perhaps even more crucial for the prey, is how long the predator can remember where they last saw their hoped-for prize. Known as working memory, this time varies greatly between species. Fortunately for a mouse hiding from a cat, it doesn't have to hold out for too long before the cat forgets about it. Experiments have suggested that over the first thirty seconds after an object

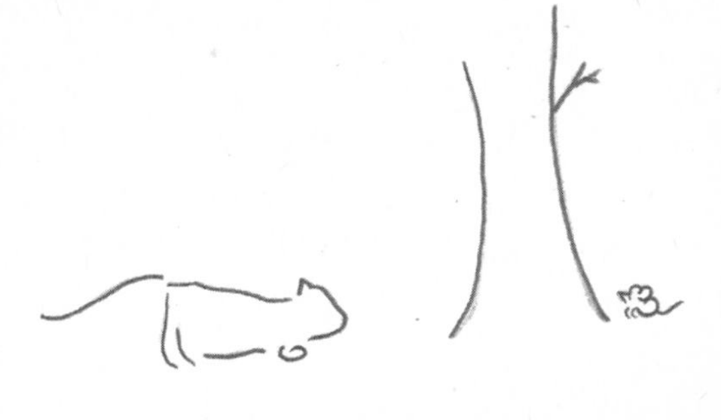

disappears from view, a cat's ability to remember the object's location wanes rapidly. After a minute, the cat's searching success is slightly better than random, which doesn't give them much hope of catching a lost mouse. Presumably, it must be more productive at this point to move on to finding a new, unsuspecting mouse rather than to waste more time looking for a mouse that knows the cat is there and is actively hiding from them.

Most modern-day pet cats are lucky enough not to have to rely on their memory and hunting prowess to survive. Meals are usually provided by their owners, and so when a cat is hungry, their visual attention will inevitably be focused on their owner and whether there is any food in the offing. Recent studies have begun to explore how much human influence might override the natural instincts of cats when it comes to searching for food.

When researcher Hitomi Chijiiwa and coworkers presented cats with a simple choice of two containers, one containing food and one without, they chose the one with food. No big surprise there, but it did show as a baseline that the cats would pick a container they knew had food inside. Other tests involved showing cats two containers, but this time there was a piece of food in each. In one test, after watching the experimenter remove and pretend to eat the food from one container, the cat was given the chance to approach and investigate whichever container they chose. In a second test using the same containers, the experimenter simply took the food from one container, held it up, and showed it to the cat before putting it back where they found it rather than "eating" the food. The cat then had the opportunity to approach the containers and again choose one.

The researchers were curious to discover whether the cats would assume that the container where food had been "eaten" in the first experiment was now empty and choose to investigate the other one instead. And in the second experiment, they wondered what effect showing and replacing the food would have on cats' subsequent choice. They found that in both experiments, the cats more often chose to investigate the container that the human had visited. However, they did this significantly more frequently when the food had been "shown" rather than "eaten." This implied that they had some idea that in the "eating" scenario the food might not be there anymore, but this knowledge was not always strong enough to overcome the influence of seeing the person handling the container.

At first sight, this might seem a strange result, suggesting that cats don't understand the concept that the food, once eaten, is not available anymore. It is possible, though, that for a domestic pet cat, the concept of the human eating their food was so unrealistic that they were checking to see what had happened to the food in the pot. They could have also been seeing if there was any more. Or they may have been so used to watching a person refill their food bowls that they assumed this was what would have happened. Either way, the authors point out that it is not necessarily a bad thing for pet cats to be drawn to the actions of humans even if, as in this case, they sometimes get misled. Other times, cats may rely on their owners to let them in or out, feed them, or rescue them from difficult situations—they obviously care more about what we are doing than we may think.

It's Rude to Stare

I was midway through a watch of the hospital courtyard. It was a gorgeous summer day, and the cats were hanging around, sunning themselves. Lunchtime had already come and gone but today an extra helping of food was unexpectedly delivered to them, courtesy of a staff member who appeared through the door at the top of the ramp. It looked like scraps of bacon on offer. Betty, resting nearest the door, was first to approach the food tray. As she sat chewing her way through a piece of the food, Tabitha crept up and crouched nearby. She watched Betty. I say watched, but it was really more of an unwavering stare. Betty, although quite obviously aware of Tabitha's staring, kept her eyes averted as long as she could bear before succumbing to the pressure. Eventually she moved to one side and Tabitha approached the tray to eat. Betty stared at her.

I recorded interactions like Tabitha and Betty's often when the colony cats were eating. Researcher Jane Dards, who studied interactions of feral cats in the dockyards of Portsmouth in the UK in the 1970s, documented similar stares between those cats. Dards named the behavior the food stare, its purpose presumably being to intimidate the cat currently feeding into surrendering the food. To me, it was always slightly reminiscent of a child intently and hopefully watching another unwrap sweets, one by one. I encountered the food stare in rescue shelters, too, when pairs of cats that had been brought in together were housed in the same pen. Even when they had the most affectionate relationship with each other, this staring often occurred around food bowls. It occurs in domestic households, too—a good reason to provide each cat with its own food bowl and to position them some distance apart.

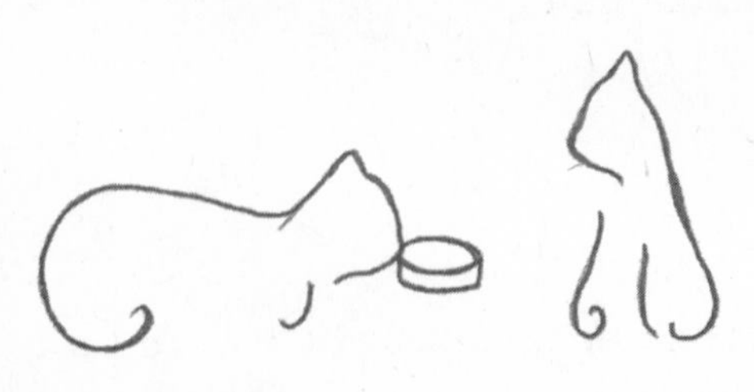

My records of interaction sequences between the colony cats often started with one cat watching another. In my early days of observations, if the recipient cat looked directly back at the initiator, I would note it as "eyes meet." As time went on, I realized this conjured up rather romantic images of cats locking eyes across a crowded room, so I changed my terminology to a more literal "watch cat initiator" followed by "watch cat recipient." Descriptions aside, I soon realized that the moment of "eyes meeting" was pivotal in cat-to-cat encounters. What followed depended very much on the relationship between the two cats, varying between a rapid avert gaze/look away, a more intriguing mutual gaze, and a full-blown staring match.

Some smaller-scale studies have investigated these different types of looking in more detail. Deborah Goodwin and John Bradshaw examined interactions between cats from the same colony and broke them down into those that included aggressive elements and those that didn't. They found that in interactions that contained aggressive behavior, there was less mutual gazing between cats than expected, given the total amount of time each one spent checking out the other. These cats were obviously watching each other covertly, looking away when the other looked at them. When the encounters contained only nonaggressive behavior, the amount of mutual gazing was as expected given the total amount of time each cat spent looking at the other. In these cases, mutual gaze appeared to be nonthreatening in nature, unlike the warning stares of hostile encounters, or the frosty "food stares" of Betty and Tabitha.

Another study by the same authors looked at staged interactions between pairs of cats, specifically to record eye contact in more detail. They found the most common sequence of behavior was for a cat to look briefly at the other and then look away, as if they were monitoring what the other cat was doing. Interestingly, this was then

often followed by the same cat either sniffing a nearby object or grooming themselves. These actions are commonly used by cats as displacement behaviors, a way of releasing tension when not sure what to do next. Cats sometimes avoided interaction following a look away. If, however, neither cat looked away, leading to a mutual gaze, then approaching and sniffing often ensued. Again, this was evidence of a different kind of more friendly mutual gazing compared with the confrontational staring of cats in an aggressive encounter.

While staring between individuals has traditionally been considered a hostile behavior among animals, these studies and those of other species indicate that in some situations, mutual gaze between individuals can reflect a more friendly relationship.

Humans act similarly. Although sometimes a hard stare is used to exert dominance, we look people in the eyes for many other reasons. We may "catch their eye" to work out how they are feeling, to check if they agree with something we're doing, or to let them know we wish to engage in an interaction. Sometimes we do this silently, in isolation from any other form of communication. Mostly, though, we make eye contact with someone as we talk to them.

There are moments during conversations in which both people are looking at each other simultaneously. This "mutual gaze" can also occur without vocal conversation, just as I observed in my colony cats. The famous sociologist Georg Simmel wrote, "The eye cannot take unless at the same time it gives," referring to how, when two people share a mutual glance, they both provide and receive information. Just as between cats, looking directly into someone's eyes may be interpreted differently according to the relationship between the two people. If the gazer knows the recipient, or is engaged in conversation with them, then it may feel friendly and encouraging. A long silent stare from an unfamiliar person, on the other hand, may be interpreted as hostile. One study revealed that we prefer mutual gazes on average to last 3.3 seconds. People do, however, have very variable preferences for eye contact, and gazes that are either too long or too short tend to make the participants feel uncomfortable. In particular, people with autism spectrum disorders are more sensitive to eye contact and tend to avoid mutual gaze with others.

Cats, too, may have a preferred length of mutual gaze before they start to feel uncomfortable, at least when it comes to cat-human interactions. One small study showed that eye contact from a human affected cats' behavior in various ways. In a familiar room but faced with an unfamiliar man, eight cats were tested for their reactions when the man either looked at the cat, waited for eye contact to be made, and then turned away, or alternatively looked at the cat, waited for eye contact, and then continued looking at the cat for a whole minute. When the man looked away after initial eye contact, the cats were more inclined to look at the man more often and for longer, perhaps monitoring the situation. However, when the man stared at them, the cats mostly hid or averted their gaze from him, apart from a minority who approached the man and got on his lap.

Looking at how pet dogs and cats interacted visually with children, researchers of one study discovered that, rather than more prolonged gazing like dogs, cats tended instead to send a mixture of gazes and shorter glances toward the children. Cats and children in the study also rarely shared mutual gazes. This more reserved eye contact by cats may be a reflection of their solitary backgrounds and less developed visual signaling system compared to dogs. Importantly, though, the authors of the study suggest cats' pattern of shorter glances may provide a more comfortable method of visual interaction with pets for people who prefer less sustained eye contact. In this study, the authors referred particularly to children with autism spectrum disorders, but the same may apply to all children and adults who generally like to avoid lengthy gazes.

Looking for Help

Much interest has grown in the field of animal behavior science over the ways in which different species utilize human gaze, in particular companion animals with whom we share our lives and homes. For a long time, the focus was on dogs who, thanks to their long, loyal relationship with humankind and their innately social nature, have developed an impressive ability to read and, in some cases, manipulate our gaze. Attention then gradually turned to cats, with researchers testing their abilities using similar techniques to the ones they had used for dogs. These tests typically consist of puzzles for cats to try to solve, with or without the help of a person, usually to gain access to a food reward. Sometimes the puzzles are solvable by the cat on its own, but others require help from a person in order to get the prize—so-called unsolvable tasks.

In unsolvable task situations, humans and some animal species, while looking for a solution, often exhibit what are known as "showing" behavior patterns. This is when they attempt to attract the attention of a person who they know can help them and lead them to the object they desire. In humans, this may consist of rapid transfer of gaze back and forth between the recipient person and the object of desire (gaze alternation), sometimes combined with pointing. Dogs, unable to point with anything other than their noses, use a combination of gaze alternation and body movements to convey their message, often running rapidly back and forth between the object and the person, looking at the human until they have their attention and then at the object. But what about cats?

An interesting study by Ádám Miklósi and coworkers compared the behavior of both cats and dogs when faced with the challenge of accessing some food that they had seen being hidden but couldn't retrieve without help. For each test, the animal's owner and the experimenter who had hidden the food were present in the room while the cat or dog tried to work out how to get to their reward. Miklósi found that the dogs looked earlier and more frequently at the humans present than did the cats. The cats did use eye contact but much less so than the dogs and persisted at solving the problem themselves, without human help, for longer than the dogs.

In a different experiment, Lingna Zhang and co-researchers looked at how cats engaged visually with a person when faced with both a task they could solve alone and an unsolvable task. They found that the cats employed different strategies depending on the task at hand. When they could not solve the task alone, they spent less time with the person and approached the treat container less often but engaged in increased amounts of gaze alternation between the container and the person, compared to the task they could solve on their own. Interestingly, their behavior also varied according to how attentive the person was. The person behaved in one of two ways—in the attentive state, they would look in the direction of the cat, thereby making themselves available for visual interaction. In the inattentive state, they looked down at a stopwatch, avoiding eye contact. The cats looked at the person sooner and more often, and approached the food container more frequently, when the person paid attention compared with when they ignored them.

A similar effect was found in a study by Lea Hudson, who presented shelter cats with an unsolvable puzzle containing a piece of food. The cats were introduced into a room containing only the puzzle and an unfamiliar person. As in Zhang's experiment, the person's behavior was either attentive (they watched the cat and reciprocated eye contact when the cat gave it) or inattentive (they turned their back to the cat and avoided eye contact). Hudson found that the cats altered their behavior according to whether the person was paying attention to them. They gazed significantly longer and more frequently at the person if they were attentive than if they were ignoring them.

Cats are obviously quite attuned to whether people are taking any notice of them, and they are capable of gaze alternation in situations where they need help to access a resource. It is interesting that they only do this in earnest if we are actually looking at them—a salient thought to bear in mind in our other day-to-day interactions with them. Humans are not always the most observant of species, a fact evidently not lost on cats, who have developed other techniques, such as meowing or rubbing around our legs, to nudge us into action when they want to interact. Paying attention to their more subtle eye signals could be a way for owners to build a stronger connection with their cats.

What Are You Looking At?

Take that moment when you are chatting with someone face-to-face. Their attention wanders from your eyes to a point in the distance somewhere over your shoulder and instinctively you turn your head to see what they are looking at. This is known as gaze following, a form of information gathering, and following a person's gaze to look at the same object results in joint attention, in which you both end up looking at the same thing. People also have another use for gaze following. When interacting with each other, especially during silent interactions, we will on occasion deliberately shift our gaze to something to indicate that it may be relevant or interesting to the onlooker. Used in this way, gaze becomes what is known as a "referential signal."

Gaze following, once thought to be a skill unique to socially advanced species such as our own, has now been found in numerous other social animal groups but, perhaps more surprisingly, in some solitary species too. In particular, researchers have become interested in whether other animals can follow referential signals given by a human.

Péter Pongrácz and coworkers from Eötvös Loránd University in Budapest, Hungary, wanted to see if cats can use our referential gaze signals as clues to work out what we are looking at and thereby gain information or a reward. To make the environment as relaxed as possible, cats were tested in their own homes. In each test, the cat was released by its owner to approach an unknown experimenter who sat slightly behind and between two pots, one of which contained hidden food. Both pots were previously smeared with a little of the hidden food to prevent the cats from selecting a pot simply by scent. The experimenter initially gained the cat's attention by one of two methods. The ostensive method involved calling the cat's name or using any other sound normally used to get their attention, while for the non-ostensive method the experimenter made a clicking sound, a noise not generally used to call cats. Once they had the cat's attention, the experimenter then moved their head and looked directly at the pot containing the prized treat. To spice things up a little, the gaze cues given by the person were of two different types. One, known as dynamic gazing, involved the person looking continuously at the correct pot right up until the cat made its choice. Another type of gazing, known as momentary gazing, was also tested. In this case, the person sent a fleeting glance toward the correct pot and then looked back at the cat. While momentary gazes are more difficult for humans to follow, Pongrácz and his team wondered if the element of movement in the gesture might make it more noticeable for a predatory species, such as the cat, that is accustomed to scanning for movement when hunting.

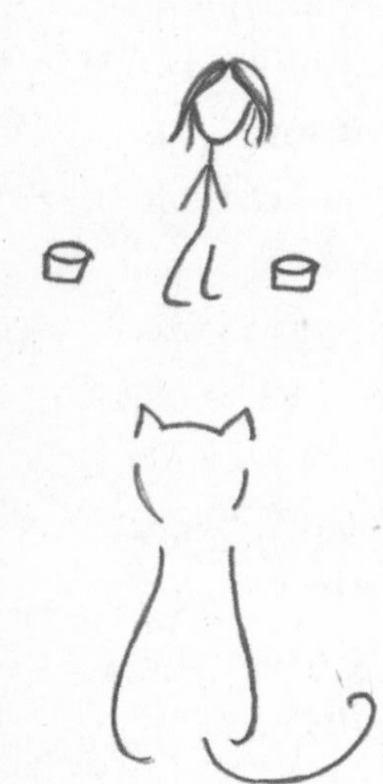

It turns out the cats were very successful at following the gaze cues given to them by the human, and an unfamiliar human at that. Overall, they made the correct choice of the food-laden pot an impressive 70 percent of the time, an accomplishment similar to that of some nonhuman primates and even matching the skills of dogs. Not only that, but they were equally good at following both the more obvious dynamic gazing and the more subtle momentary version. Interestingly, the use of ostensive versus non-ostensive cues did not improve a cat's chances of success at the task in hand, a result different from that found in dogs. The more familiar ostensive vocal prompt did, however, make the cats engage in eye contact faster than if a non-ostensive noise was used.

Although far from complete, a picture of how cats utilize human gaze is beginning to emerge from all these studies. In the domestic environment at least, cats, far from being aloof and unreceptive to humans looking at them, have learned to use this method of communication despite its being instinctively quite unnatural to them. They have learned to look at us when faced with a human-made unsolvable problem and to follow our gaze to gain rewards. These situations are of course quite unnatural in themselves, but the ever-adaptable cat will take an opportunity whenever it arises. The unique experiences and relationships domestic cats have with owners and other people may result in individuals having different abilities to follow people's gazes. The question remains as to whether each cat learns to follow a human gaze during its lifetime or whether it is an innate trait that has evolved within the species over the course of domestication. To truly investigate this would require studying how socialized wildcats might respond in similar test situations, as yet not attempted.

Getting the Point

As well as establishing that cats can follow a referential signal from us in the form of a gaze, researchers are interested in whether our pet cats can understand what we mean when they see us perform other referential signals.

Take the human gesture of pointing. By the age of one, we begin to point a finger to indicate something we want (imperative pointing) or something we want to show to others (declarative pointing). As pointing is so important to humans as a communicative device, scientists have tested numerous animal species on their ability to follow our pointing cues toward a target object. No one has ever seen a cat point or even wave a paw in the general direction of anything. Understanding what pointing means is therefore a fairly tall order for an animal without a pointing digit of any kind. On top of that, many domestic pet cats are accustomed to people holding their hand out toward them for them to sniff as they greet them. The cat's instinct in this scenario is usually to approach the hand, not to look at where it is pointing.

Despite this, in the same series of experiments in which they tested gaze following, Ádám Miklósi and co-researchers also discovered that both cats and dogs were able to follow the pointing cues given by a person when indicating which of two bowls contained a hidden food reward. They could do this whether the pointing finger was close to the bowl (4 to 8 inches away) or farther away (27.5 to 31.5 inches) and were successful whether the signal was prolonged or fleeting (just one second) in duration. Cats were as good as dogs at following the pointing, an interesting result given the reputation of dogs for being more tuned in to what people are doing. Perhaps even more surprisingly, cats were at least as good at the task as some nonhuman primates, who do have the ability, although not necessarily a natural tendency, to point. This impressive ability may be the result of cats' association with people—they have adjusted not only their own language but also their understanding and use of uniquely human signals, enabling better communication. That there is often some tasty food available at the end of the pointing finger has probably helped the learning curve considerably.

Péter Pongrácz and co-researchers explored human-cat pointing a little further as part of a survey of cat owners in Hungary. In this case, "pointing" referred to several visual cues, not only hand signals but also referential cues given with head turns and gazing as previously explored. They found some interesting effects. With respect to playful interactions between owners and cats, when the cats initiated the play session, the owners used referential cues less frequently. The authors suggest this might mean that owners who are more tuned in to their cat's signals direct them less often, whereas those who pay less attention to their cat's signals may try to instruct them more with referential signals. Alternatively, owners may be more inclined to use such gestures when they are the initiators of the game in order to encourage interactive play. Owners who stated that their cat used combinations of visual, tactile, and vocal methods of interaction tended to use referential signals with their cats more often than those whose cats used only a single mode of communication. Quite who is influencing whom in this scenario is tricky to decipher: Did these cats respond to such communications from their owners or vice versa?

In the Blink of an Eye

I crept into the cat pen as soft-footedly as I could. From the box on the shelf at the back there was a distinct shuffling noise. I stood on tiptoe and looked in. All I could see was a pair of flattened ears and the top half of a frightened-looking furry face. Two saucer-like eyes looked back at me over the edge of the box. That's as much as I saw of Minnie for the first week or so

after she came into the shelter, and I had all but given up hope of seeing much more of her. I decided to try a new tactic, one I had recently heard about—blinking really slowly. I was a little dubious. But as a young student, I realized I had much to learn and little to lose as long as I didn't get too close or try to touch her. I teetered once again on my tippy-toes, peered into Minnie's box, and, feeling slightly foolish, tried slowly blinking and then squinting so my eyes appeared half-shut. It was hard to actually see much through my half-closed eyes, but to my astonishment the saucers looking back at me performed a slow blink and then stayed half-shut in response to mine. Did I imagine it? Realizing my own eyes

had flung wide in surprise, I repeated the slow-blink move. So did Minnie.

For a while, everywhere I went after that I would experiment with the "slow-blink" method of communication with cats. I tried it with cats in rescue centers, cats in my own and other people's homes, and even cats I met on the street. It became obvious to me that it was a familiar signal to many of these cats as they politely responded in kind to my somewhat clumsy attempts at wooing them. To my disappointment, this subtle slow-blinking display was already known anecdotally to many of the people I excitedly described it to—those who worked with cats such as vets, vet nurses, and rescue center workers, and current and previous owners of cats. Nevertheless, I was thrilled; it felt like I had suddenly been allowed into a secret signal society.

I quizzed people about when they mostly used their slow blinking and got answers like "Oh, you know, when we've been kind of looking at each other across the room for a bit too long" from owners, and "When cats first come in and they are all crouched and scared" from rescue shelter workers. This last description reminded me of Minnie.

A survey of cat owners carried out by the British cat rescue organization Cats Protection in 2013 revealed that 69 percent of owners associated the slow eye blink in a cat with its feeling relaxed in their company. Although anecdotally well known, the detail of these mysterious cat-human eye-narrowing conversations was never investigated scientifically until Dr. Tasmin Humphrey and co-researchers at the University of Portsmouth put it to the test. In their paper they describe how the cat slow-blink sequence includes a series of half blinks in which the eyelids only half close;

then they may stay in this half-closed state for a while (described as eye narrowing) or close completely.

The team looked in more detail at how everyday cat-human eye-blinking encounters might work by video-recording and analyzing staged encounters between owners and their cats in their home environments. In one test condition, the owner was asked to engage the cat's attention and then perform slow eye blinks at it. In the other, the owner was present but ignored the cat and did not slow-blink at it at all. They found that the cats were significantly more likely to perform slow-blink behavior toward their owners if the owner had slow-blinked at them. So slow eye blinking wasn't just a random occurrence between owners and cats.

The researchers then went on to see cats' reactions to an unknown person when that person looked at them and performed slow eye blinks, compared with when they kept a neutral face without making direct eye contact. They found that not only did the cats respond with more slow blinks themselves, but they also preferred to approach the stranger after they had slow-blinked compared with when they merely kept a neutral expression. This supported the generally held assumption that slow eye blinking puts cats at ease, encouraging them to interact further with both familiar and unfamiliar people.

But what about cats that aren't in a place where they feel relaxed—cats like Minnie that suddenly find themselves deposited in the unfamiliar environment of a rescue shelter? Add to this the stress of prospective new owners peering into each pen and staring at the resident cat for some time as they appraise them. Just like Minnie, many cats at shelters also appear to use the slow-blink response when faced with such stressful situations. A second study by Humphrey and co-researchers looked at slow-blinking reactions by cats housed in shelters rather than in homes, and their reactions to an unfamiliar person giving them slow blinks. Again, slow blinking by the human experimenter resulted in higher rates of slow-blink responses from the cats. The cats in these tests varied greatly in the levels of anxiety they exhibited with their living situation, and yet the less anxious ones did not slow-blink more frequently than more anxious individuals. In fact, there was a slight although nonsignificant tendency for the more nervous cats to spend more time producing slow-blink sequences in response to human slow blinks. This suggests that slow blinking may serve different purposes according to the circumstances. When the setting is calm and friendly and the cat is relaxed, it may serve an affiliative function. But in a frightened cat, it may be a form of submission, to disperse tension. Such dual-purpose signals are not uncommon in animal species, since the ultimate objective in any social group is to both avoid confrontation and increase harmony among its members. Cats themselves have another such signal in the form of mutual grooming, discussed in chapter five.

The slow blink is another of those behaviors for which it's hard to know, did cats learn to do it to each other first? Or did they develop it with people and then find it worked in cat society too? Prolonged unbroken staring in cats can be a hostile behavior, likely to progress to a more intense, aggressive interaction unless one or both cats look away. It is possible in cat-cat situations that the slow blink serves as an alternative to looking away—a way of softening the gaze and indicating nonhostile intentions.

As Humphrey and coworkers point out, it's not just cats that display this eye-narrowing routine—canids, horses, and cows do it, and humans too. Although we are able to artificially slow-blink at cats, the behavior is very reminiscent of a natural eye-narrowing behavior displayed by humans. We do this unwittingly when we perform what is known as a Duchenne smile, named after the

French neurologist Guillaume-Benjamin-Amand Duchenne de Boulogne, who discovered it in 1862. Duchenne was one of the earliest scientists to study the nerves and muscles of the human face and how they interact to produce expressions. One of his conclusions was that, while all human smiles require contraction of the zygomaticus major muscles in order to raise the corners of the mouth, a genuine happy smile also results in the contraction of the muscles around the eyes, the orbiculares oculi. The result is a visible narrowing of the eyes and a wrinkling of the outer corners of the eyes into "crow's feet."

It has since been shown experimentally that many people can imitate this eye-wrinkling effect quite convincingly, for other humans as well as for cats. Perhaps our eye narrowing creates, as Duchenne himself put it, an "agreeable impression" for cats, to which they feel inclined to reply. As it has evolved, our ability to replicate eye scrunching, and cats' ability to mirror it, has created an unexpected and charming channel of visual communication between humans and cats—the opportunity to smile at each other with our eyes.

CHAPTER 7

THE PERSONALITY PUZZLE

There are no ordinary cats.

—Colette

I currently share my home with two wonderful cats—Bootsy and Smudge, sisters adopted from a shelter together at eight weeks old. They were inseparable as kittens, racing around the house in play and curled up together in sleep—tiny bundles that would fit in your hand. Now, at the mellow age of fifteen, as is so often the way with adult siblings (feline ones, at least), they avoid each other whenever they can. They came from the same mother and litter, and have shared the same home, love, and attention. And yet they are as different as chalk and cheese.

The enchanting children's book *Six-Dinner Sid* by Inga Moore, a must-read for any cat owner, describes Smudge perfectly. On a summer day she leaves the house first thing in the morning (after breakfast, obviously) and comes back as evening falls in response to me sweetly calling several times and eventually yelling several more times from the back door. In between, she will have visited any number of houses in the surrounding streets. She returns smelling of the perfume, food, and homes of other people. On less adventurous days she posts herself on the front garden wall and solicits attention from people who pass by—everyone stops to chat to her. One day I was getting out of the car in the driveway and called over to her, only for a passerby to say, "Oh—is she *your* cat? She's always in our house so we thought she was a stray." I glanced at Smudge's shining coat and well-padded little body. Seriously? A stray? Outwardly I smiled, explained that she always loses her collar, and politely (I hope) implored them not to feed her when she next visited them. We chatted further, and I asked if our other cat had ever visited them too. "You have another cat?" they replied.

Our neighbors never see Bootsy, the home lover. She spends her days mostly indoors, basking in a pool of sunshine somewhere or, in summer and autumn, scooting around the back garden, chasing flies or leaves. We also had, until very recently, a golden retriever. Alfie was the softest, goofiest, greediest, and possibly smelliest dog you could ever meet. Bootsy loved him, though—she would rub herself all around his body, so much bigger than hers, and then snuggle up with him in his bed at night. Smudge could barely hide her disdain for Alfie—one look from her would send him running to his bed.

How did two littermates end up so completely different?

Personality—at least, human personality—has intrigued philosophers and scientists for over two thousand years. Our personality affects the way we approach life, how we view the world, and, importantly, how we communicate with one another. The Greek philosopher Hippocrates (400 BCE) believed that people's personality characteristics were influenced by four fluids, or "humors," within the body, an idea Galen (AD 140) later expanded on. A choleric humor (generally bold and ambitious) was dictated by yellow bile; melancholic humor (more reserved and anxious) resulted from black bile; sanguine humor (optimistic and cheerful) came from red blood; while phlegmatic humor (calm and thoughtful) originated from white phlegm.

Personality theories and research have changed dramatically through the years, and luckily, we no longer associate character with bodily fluids. One of the most heavily researched aspects of human personality today is what is known as the shy-bold continuum. This refers to how people react to new or risky situations, with bolder individuals taking the most risks and shier ones avoiding them. The tendency to be shy or bold was the subject of a groundbreaking study on children that began in the 1970s by Jerome Kagan. Kagan's work showed that while shyness was an inherited trait, it was not set in stone and that, in many cases, environmental conditions could reduce shyness as the child developed. Bold children were less likely to show changes in their boldness as they grew older, presumably because they were not encouraged, as were shy children, to become more confident.

While interesting, looking at the shy-bold spectrum considers only one aspect of personality. Researchers began looking for a way to describe a broader overall picture of personality, measured in more than one dimension. The eventual result was what is known as the five-factor model of personality, often referred to as the Big Five. Widely considered to be an accurate and repeatable personality test for people, the Big Five uses a questionnaire to score individuals on five aspects: Openness to experience,

Conscientiousness, Extraversion, Agreeableness, and Neuroticism. (Neuroticism refers to a disposition toward negative traits, encompassing depression, vulnerability, irritability, moodiness, anxiety, and shyness.) Scores for each of the five dimensions are made on a scale so that a person is not categorized, for example, as simply extraverted or not, but instead appears somewhere on a continuum between high and low extremes. Particularly assertive and sociable people receive high scores on the Extraversion scale, very quiet and reserved people score toward the lower end of it, and most people will score somewhere in between. The combined scores for each of the five factors form the personality profile for the person taking the test.

Animal Personalities: "To Boldly Go"

Do cats and other animals have personalities too? Ask any cat owner and they will tell you, "Why, yes, of course!" before regaling you at great length with the important details thereof (see my own aforementioned account of Bootsy and Smudge). Yet, despite around two thousand years of thought on the topic in humans, for many years the consensus in the scientific world was that for nonhuman animals, domestic or wild, personality did not exist. Individual variation was simply regarded as background "noise" during studies of animal populations. To suggest that animals had personalities raised cries of anthropomorphism of the worst degree, an accusation guaranteed to strike horror into the minds of serious behavior scientists.

However, as well as favoring individuals with certain physical traits, natural selection also works on how animals respond in different situations. Individual variation in behavior is therefore a key factor in the study of animal communication, ecology, cognition, and evolution. Acknowledging this, scientists slowly began to pay more attention to the differences between individuals in their studies. One of the earliest scientists to introduce the concept of different animal personalities was the Russian scientist Ivan Pavlov, famous for his studies of conditioned responses in dogs in the late nineteenth century. He came up with four personality types in his dogs, similar to the ancient Greek system devised for humans: Excitable (Choleric), Lively (Sanguine), Quiet (Phlegmatic), and Inhibited (Melancholic).

Gradually the study of personality in animals became an acceptable science. In species from apple snails to marmoset monkeys, researchers have explored different traits within individuals and between populations to see how they affect communication, how they aid survival, how they are inherited, and how they vary in different locations. In a bid to avoid any hint of anthropomorphism, scientists have often dodged the word "personality" itself (it does, after all, contain the word "person"), instead using terms such as "coping styles," "behavioral styles," "behavioral syndromes" or, if they are feeling very brave, "temperament" (although technically "temperament" refers more to the inherited part of personality).

Just as with humans, a particularly well-studied aspect of animal personality has been the spectrum of shy to bold personalities, a feature that seems to crop up in every animal population. Bootsy and Smudge are classic examples of this—bold Smudge, who wanders confidently into my neighbors' houses, and shy, retiring Bootsy, who sticks to her small modest patch at home. Researchers have wondered how two such opposing character types can succeed and perpetuate in a population from one generation to the next.

Because many species nowadays have had to adapt to the ever-greater proximity of humans, studies have examined how human activity, along with its environmental consequences (the anthropogenic effect), has impacted the behavior and personalities of various animal populations. The growth and spread of urban areas has created new niches, ripe for exploration by animals that normally inhabit the rural areas surrounding them. Availability of novel food sources is a big draw for wild species constantly on the lookout for the next meal, and an animal's ability to adjust to new challenges dictates how successful it will be in colonizing these environments. Studies indicate that, perhaps not surprisingly, it is the bold or proactive individuals in populations that dominate such new urban niches—they are often referred to as "urban adapters." For example, Melanie Dammhahn and coworkers investigated populations of striped field mice (*Apodemus agrarius*) in four urban and five rural locations in Germany who exhibited different levels of urbanization and environmental disturbance caused by humans. They found that urban mice were bolder, more explorative, and more flexible in some of their behavior than their shier, rural cousins.

Studies like this inevitably raise a question: If bold, proactive animals gain first access to new food sources, new places to shelter, and, as a result, potentially new opportunities to reproduce successfully, why are there still shy individuals within a population? The simple reason is that while boldness may bring benefits, it also comes with risks. In urban areas, for example, along with new feeding opportunities come new predators and parasites, road and traffic hazards, and polluted air, soil, and water.

Ancestors of the modern domestic cat, the opportunistic wildcats of the Fertile Crescent ten thousand years ago described in chapter one, might be regarded as the original feline urban adapters. They, too, probably had a mixture of bold and shy personalities among each local population. The early human settlements that sprang up within their natural territories were a primitive prototype for the anthropogenic world we see today.

For sure the bolder, braver cats back then would have been the ones who first crept into these primitive villages to see what they could scavenge.

But these feline pioneers probably met with mixed reactions. Like them, the newly settled farmers would have been opportunistic, and many wildcats may have ended up as dinner for a farmer's family rather than securing their own much-hoped-for meal. Others may more successfully have wooed the farmers with their rodent-controlling talents and been allowed to hang around. What, then, of the shy cats? Shier wildcats, taking their time to scout out and assess the newfound resources before approaching, may have avoided the risk of predation by humans and other carnivores such as wild dogs, thereby surviving longer and reproducing for a longer time than the bolder individuals. Both tactics could succeed but in different ways.

As chapter one follows in detail, the fortune of the cat at the hands of humankind has been something of a roller coaster—at any one point in this journey, a bold or shy personality could be an advantage or a detriment. In ancient Egypt, cats were so revered they could probably have thrived whether bold, shy, or anywhere in between. Cats of the Middle Ages, on the other hand, may well have benefited from a shier demeanor. Persecuted by humankind, keeping a low profile was their best policy. Bolder male cats in modern-day feral colonies have been shown to have increased reproductive success. They pay for this, however, with an increased risk of contracting feline immunodeficiency virus (FIV), typically transmitted via bite wounds when tomcats fight. Thus, although reproductively efficient, they may live shorter lives than shier individuals.

The ongoing survival of cats, and indeed many species, seems to rely on their maintenance of a mix of shy and bold personality types that will be selected for or against according to the prevailing local conditions. However, even the shyness-versus-boldness spectrum may not be as straightforward as it seems. Looking at Bootsy and Smudge a bit closer, it turns out that "shy" little Bootsy is actually very confident with human visitors to our house and, as we saw, was also very sociable with our dog Alfie. Smudge studiously ignores most visitors when she is at home herself and had no time for Alfie either. Just like people, cats have complex personalities.

Cat Personality

Aloof, independent, sly, timid, affectionate, spiteful, intelligent, playful, inquisitive, sneaky, confident, shy, suspicious, enigmatic. Cats throughout history have been labeled with an impressive number of personality traits by the subjective observations of cat haters and cat lovers alike. Finding a more objective and scientific way of describing cats' personalities and their differences from one another has proved equally as challenging as for human personalities. Attempts began in earnest in 1986, with a study that identified three basic dimensions of cat personality, defined by the authors as Alert, Sociable, and Equable. Later investigations built on this, using a variety of methods and terminology to isolate between four and seven personality dimensions. Assessments of cats in these studies mostly focused on how they behaved toward humans rather than other cats.

A breakthrough came in 2017 with the biggest and most comprehensive study of cat personality to date. This analyzed data on over 2,800 cats from questionnaires completed by cat owners across South Australia and New Zealand. From these, Carla Litchfield and co-researchers came up with five dimensions of personality, made up of fifty-two individual traits that, importantly, included behaviors directed at other cats as well as at humans. Known as the Feline Five, these dimensions bear some intriguing resemblances to the Big Five originally elucidated for humans. Cats were found to have some personality dimensions similar to those of humans: Agreeableness (also described as Friendliness for cats), Extraversion (Outgoingness), and Neuroticism (Skittishness). Most owners would probably not be surprised to learn that cats lack the human Conscientiousness dimension (apparently only found in humans, chimpanzees, and gorillas). They also lack the Openness dimension. Instead of these, cats have a dimension named Impulsiveness (or Spontaneity), and another that is also found in other nonhuman animal species, Dominance, reflecting their tendency to bully or show aggression toward other cats at one extreme or to act submissively and amicably toward them at the other.

The individual traits that make up the five feline personality dimensions include qualities such as insecure, anxious, suspicious, and shy, which are encompassed by Neuroticism (Skittishness); and affectionate, friendly to people, and gentle, which are grouped within Agreeableness (Friendliness). The fifty-two possible traits also include some less common but wonderfully perceptive options. My particular favorites, having met some cats with these enchanting qualities over the years, are reckless, aimless, quitting, and clumsy.

Just as with the human Big Five, each of the dimensions in the Feline Five is represented by a low-to-high scale, with the behavioral extremes at either end. Cats receive a score for each dimension based on their owners' answers. They generally score somewhere in the middle for a few dimensions, with perhaps a high or low score on one or two of the others. For example, Smudge, just like the proverbial curious cat, would score high on the Extraversion dimension, high on Agreeableness, and low on

The Five-Factor Scale of Domestic Cat Personality

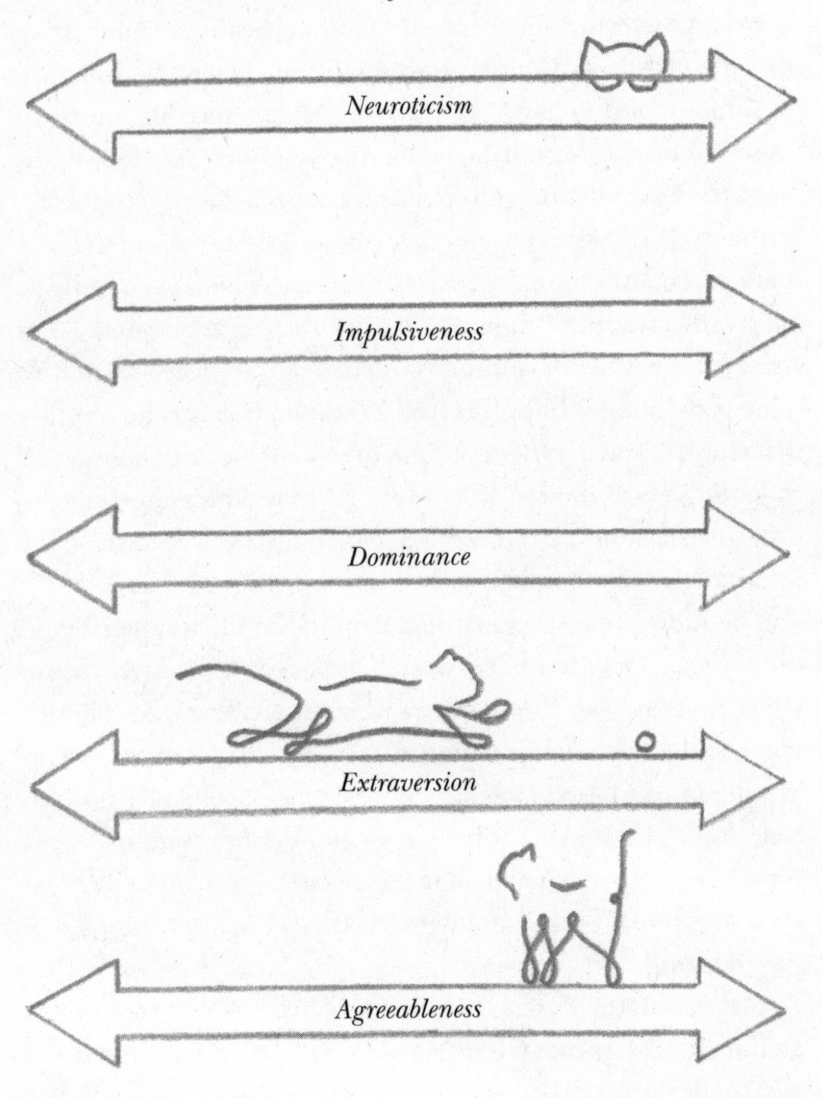

the Neuroticism scale. My jumpier Bootsy would score higher on Neuroticism and Impulsiveness, but she, like Smudge, loves people, so would score high on Agreeableness too.

The Feline Five test gives an owner the opportunity to look at their cat's temperament from a much wider perspective rather than simply in terms of single traits such as "my cat is very curious" or "my cat is friendly." A cat might be curious as well as outgoing and friendly, or curious but more solitary in nature.

The personality profile also spotlights potential ways to improve an individual cat's welfare. For example, cats that score high on the Neuroticism (Skittishness) scale are often stressed, so creating more places to which they can retreat and hide in the home can make them feel more secure. Active, curious cats, scoring high on Extraversion, sometimes lack stimulation, particularly if they are indoor-only pets. Providing them with extra toys and new things to explore, such as empty cardboard boxes, may help alleviate boredom. Cats that score high on Agreeableness (Friendliness) may benefit from increased attention from their owners, and they will be most receptive to interactive play and petting.

In addition, looking at cats' profiles may help when considering whether to introduce another cat to a household, enabling the selection of cats with temperaments that complement each other rather than ones that clash. For example, two cats scoring low on Dominance are less likely to quarrel over shared resources. Some individuals—like those who score high on Dominance or particularly low on Friendliness—may actually be better off as the only cat in a household.

What decides what type of personality a cat will have? What determines whether they will be a cat that likes to purr, rub, and meow recklessly at people or whether they will be more reserved, independent, suspicious, quiet, or, heaven forbid, aimless? Scientists have worked hard on this question, so crucial is it to the success of any cat-human relationship. As ever with cats, the answer is not a simple one; just as with humans, it seems to be a mixture of genetic and environmental factors, the age-old nature versus nurture conundrum.

Environmental Contributions to Cat Personality

Most people, if asked, would admit that they would prefer a friendly cat as a pet rather than an aloof or shy one. Friendliness has been, perhaps not surprisingly, the most heavily researched aspect of cat personality. So what exactly makes a cat friendly toward humans? How do you choose a kitten that will turn out to be friendly as an adult? The large number of feral cats that lurk around our towns and cities, avoiding contact with people at all costs, are a sharp reminder that cats, even though domesticated, are not actually preprogrammed to be friendly to humans. They can manage quite well all by themselves without gracing us with the sweet meows, raised tails, and head rubs that we have come to expect and enjoy. To become friendly to us, they must learn that interaction with people is a good thing.

There is a period early on in every animal's life when they are open to making social attachments to both their own and other species, a process known as socialization. The young of some species, typically birds, are born already well developed (known as precocial) and tend to follow the first thing they see that moves after they are born. Ideally this will be their mother, but it can also be a human or other animal. This phenomenon was described as "imprinting" by the Austrian ethologist Konrad Lorenz, who was famously followed around by a gaggle of goslings he had intercepted at birth.

For slower-developing species, including dogs and cats, there is a longer window of time, still while they are very young, for them to learn which animals are good to interact with. Puppies and kittens within a litter feed and play together as they develop, naturally learning from one another and from their mother how to interact with their own species. Learning how to interact with people, however, requires a more considered exposure to humans at an early age. Research in the 1950s and '60s identified this so-called sensitive period of socialization for dogs to be between three and twelve weeks old.

Although various scientists suggested different ideas of when this sensitive period might be for kittens, for a long time no actual research was conducted to prove it either way. Most people assumed that cats were similar to dogs, and that petting and handling kittens when they went off to new homes around eight weeks old was fine for socializing and familiarizing them with people. Then, in the 1980s, a groundbreaking discovery was made by researcher Eileen Karsh. Karsh designed a series of experiments to try to pinpoint more precisely the sensitive period of socialization for kittens. Based on the premise (from studies on puppies) that seven weeks was around the midpoint of this important period, she split her study cats into three groups. One group received petting for fifteen minutes a day from three to fourteen weeks of age, the second group had the same petting but from seven to fourteen weeks of age, while the third group received no handling at all before the age of fourteen weeks. Once the kittens reached fourteen weeks old, Karsh began to measure each one's friendliness to humans by assessing their willingness to both approach a person and to be held by them. These tests were repeated every two to four weeks until the kittens were at least a year old.

As expected, the kittens that had received no handling were significantly less friendly than the group that had been handled from three weeks old. What was surprising, though, was the similarity between the group whose handling started at seven weeks and the completely unhandled group. The former behaved almost as if they had had no handling at all. Karsh's work was a revelation—starting to pet kittens at seven weeks old was simply too late to socialize them to humans. In her subsequent experiments, Karsh identified that the precise sensitive period for the socialization of kittens is between two and seven weeks old.

This discovery meant that many kittens had been leaving their mothers—and sadly, some still do—without the early socialization necessary to cope with living with people. With such a small window of time in which to experience positive interaction with people, it is no wonder that some cats end up fearful of and hostile toward humans.

Karsh went on to explore kitten socialization in more detail, discovering that the amount of handling that kittens receive at this time is also important. She compared the behavior of cats that as young kittens had been handled for just fifteen minutes a day to those that had received forty minutes of handling each day, finding that the latter were faster to approach people and could be held for longer than the former. Plenty of gentle handling is therefore advantageous for increasing kitten sociability, up to about an hour a day, beyond which any further increase in handling has little effect on a kitten's subsequent friendliness. Karsh also found that cats' sociability was affected by the number of people who handle them—those handled as kittens by more than one person seemed to be more sociable with people in general, compared with cats that had experienced only one handler as kittens.

Outside of experimental conditions, for free-ranging feral or stray cats, or sometimes even inside a caring home, the behavior of the mother cat toward people can have a profound effect on the subsequent attitudes of her kittens. Friendly mother cats may produce friendly kittens partly because she will, through her own response to people, encourage her kittens to interact with humans too. An unfriendly mother cat may hide her kittens away and actively prevent them from being handled by people, thereby potentially missing the crucial human socialization window. Importantly, it is necessary only for socialization to *start* in this two- to seven-week period—once it has begun, a kitten can continue to build upon its positive early experiences beyond the seven-week point.

Even with a friendly mother and plenty of early handling by humans, some cats seem to end up less friendly toward people than others. This puzzled early cat ethologists, who began searching for a reason why.

The Father Factor

As well as influencing her kittens' behavior by her own responses to people, a mother cat will also contribute genetically to her kittens' personality, passing on aspects of her temperament. It is difficult, however, to separate the two. Male cats, on the other hand, generally have a paws-off approach to kitten rearing. Most, in fact, will never even see their progeny apart from happening to pass them on the street. Any paternal influence on kitten personality is almost guaranteed to be solely via the tomcat's genes rather than from any day-to-day direct influence on their behavior.

Researching cat paternity as part of her doctoral studies in Cambridge, UK, in the 1990s, Sandra McCune, working with twelve litters of kittens born to two known fathers, made a startling discovery. One of the fathers was "friendly"—he would greet an approaching person with his tail raised, rub around the person, and knead them with his paws. The other father was most definitely "unfriendly"—on seeing a person approach, he would avoid making eye contact and stay at the back of his pen with ears and body flattened, his tail tucked under him. The friendly and unfriendly males each fathered half of the litters. In an added dimension to the study, half of each male's offspring received early handling (five hours every week between two and twelve weeks of age) by humans, while the other half received no socialization. So, there were four groups in total: socialized, friendly fathered; unsocialized, friendly fathered; socialized, unfriendly fathered; and unsocialized, unfriendly fathered. When the kittens reached a year old, they were all tested in the same way to gauge their reaction to a familiar person, a stranger, and a novel object. Not surprisingly, given what we know from Karsh's experiments, all the cats socialized as kittens turned out friendlier to people than the unsocialized ones. Interestingly, McCune also discovered a separate genetic effect, with the friendly father producing friendlier kittens than the unfriendly father. There was a kind of compound effect, too, so that the socialized, friendly fathered kittens were the friendliest combination of all the possible groups.

However, perhaps the most intriguing result was the kittens' varied responses to a novel object. Whether a cat had been socialized as a kitten had no effect on whether they would approach a novel object, but those with the friendly father approached and explored the object more quickly than did those with the unfriendly father. These kittens had inherited the tendency to approach new things in general, whether they were humans or simple objects. McCune realized that this inherited characteristic wasn't actually "friendliness" and described it instead as "boldness." For the first time, someone had managed to tease apart the effects of genetics and environment on cat personality. Told simply, bold fathers produce bold kittens that will approach things. Socialization affects cats' responses only to people. This socialization itself can, in turn, be affected by kittens' boldness and may result in them becoming more sociable faster. However,

provided they receive enough handling at the right time, shy kittens may turn out equally friendly as bolder kittens in the long run. In this way, the shy versus bold tendencies of cats mirror the ones seen in humans. Just like Kagan's study on children showed, genetically inherited shyness in cats can be overcome with the right environmental conditions.

McCune's experiments showed that knowing the personality of the father cat goes some of the way toward predicting how friendly his kittens will be. Within the controlled breeding conditions of a scientific study or in the home of a breeder of pedigree cats, this is fairly straightforward. Outside of these situations, though, where a female cat is free to wander outdoors and choose her own mate, the story is, as ever, a little complicated.

Owners of pregnant female cats often quite logically assume that the father of her kittens is the local tomcat that passes through their garden. Sometimes, if, say, he is a ginger tom and the kittens have predominantly ginger or tortoiseshell coats, their assumption is likely to be correct. A fascinating study by Ludovic Say and colleagues, however, discovered that the story of kitten paternity is far more complex than that.

In areas where fewer cats live and, consequently, unneutered female cats are harder to find, such as in a rural location, intact males will often have large territories that encompass those of several females. Most of the time an unneutered female cat will be uninterested in the males, except for when she becomes sexually receptive, known as being in estrus. Then it is a whole different kettle of fish. In estrus, the female very actively invites males to mate with her. And not just once—she will mate many times and with various males if they are available. There is a practical reason for this, because ovulation in cats is induced rather than spontaneous, and often multiple mating encounters are required to induce it. When a female comes into estrus, the male whose territory overlaps hers may be able to achieve the majority of the matings by preventing other wandering males from gaining access to her.

Where there are many cats within a small area, such as in urban or suburban environments, the picture is rather different. If in this area several females come into estrus at the same time, it is difficult for any one male, however large or aggressive, to secure all the possible matings. In this scenario, gatherings of males may develop around females in estrus, and mating opportunities are shared. Interestingly, this also provides a chance for both bold and shy males to contribute to the future gene pool of the population.

Using DNA samples, Say and colleagues were able to work out the paternity of eighty-three litters of kittens, fifty-two of them born in an urban environment and thirty-one in a rural location. They discovered that 77 percent of the urban litters contained a mixture of fathers (one litter had five different fathers). Among the rural litters, however, only 13 percent had more than one father. Without testing their DNA, it is therefore almost impossible to work out which male fathered any one of the kittens in a given litter in an urban area. It may be slightly easier in a rural setting, but who knows which other tomcats might have snuck into the equation?

All in all, given the different contributions of early human contact, maternal influence, maternal and paternal genetics, and the often unpredictable nature of paternity, choosing a kitten will always be something of a lottery in terms of its personality. People have often asked me about choosing kittens—how can they tell what they will be like when they are older? The answer is really that you can't predict too precisely when they are young. Young kittens as they scoot around in play are, just like human infants, still developing their personalities. Certain traits appear to be established fairly early on—for example, in a longitudinal study of litters of kittens, John Bradshaw and Sarah Lowe found little change in the cats' boldness over the first two years in their new homes. Other aspects of their personalities may develop more slowly as cats' experience of the world around them grows. It is often the case that two littermates, like Bootsy and Smudge, so bonded as kittens, grow apart as they reach maturity and their preferences and personalities become more set.

The most important element is to ensure cats have been handled by plenty of people from the age of two weeks onward and that they have encountered as many domestic sounds and sights as possible at an early age. Many people decide instead to opt for an adult cat, combining the satisfaction of giving a rescued cat a second chance with the ability to see the cat's personality more fully formed. For adult cats, though, finding a home for the second time around can be hard—they may need to do some serious persuasion to talk their way out of the rescue shelters. What do people look for when they choose a cat? What do cats look for when they choose a person?

Coats of Many Colors

"So, what sort of cat are you looking for?" I asked the lady and gentleman looking to find their next cat from the rescue shelter. "Oh, we're not fussy. The only important thing is the personality," the lady replied. "We want a youngish, friendly cat, one that will greet us when we come home and sit on our laps in the evening." "Great—I have just the cat for you," I announced, happily marching them down to the cattery to meet their perfect new furry companion. Pebbles did not disappoint—well, she didn't disappoint me, at least. As I opened the door to her pen she jumped down from her box on the shelf where she had been sleeping and, stretching lazily, raised her tail and approached the lady, who was now crouching on the floor, climbed onto her lap and curled up, purring like a motorboat. I smiled broadly, waiting for the magic to happen. It didn't. The lady stood up, deposited Pebbles back on the floor, and announced, "No we can't have this one—she's a tortoiseshell; they aren't very friendly." "But . . ."

Give a prospective cat owner a questionnaire and ask them what they look for in a cat and they will choose personality above all else. Give them a row of different-colored and -patterned cats to choose from and something else kicks in. Although people might place personality over coat color as the important factor, there seems to be a perception that, for some coat colors, at least, the two aspects are linked. The association of different coat colors with particular personality types is ages old and, as with many traditions, has gained ground over the years.

Tortoiseshell and calico cats, for example, despite having attractive markings, have been much maligned regarding their apparent personality traits. As far back as 1895, Dr. Rush Shippen Huidekoper's book on cats described the tortoiseshell cat as "not over-affectionate, and sometimes even sinister and most ill-tempered in its disposition." The phrases "naughty tortie" and "tortie-tude," used commonly nowadays in the world of cats, have only served to increase this trend. In a more modern survey of people's impressions of different coat color associations, conducted by Mikel Delgado and colleagues, tortoiseshell and calico (also described as tricolor cats) still scored highly on aloofness and intolerance and low on friendliness.

Somewhat bizarrely, pure ginger cats are much more popular personality-wise. Huidekoper described them as "good-natured domestic cats." His opinion seems to have held on—in Delgado and colleagues' study, ginger cats were rated as relatively low in shyness and aloofness and high in friendliness, in comparison to cats with other coat colors. In a later survey of cat owners, ginger cats received the highest scores for friendliness, calmness, and trainability. They were always hugely popular in the rescue shelter where I worked—people would regularly ask if we had any ginger cats available for adoption.

Huidekoper had strong opinions on some other personality characteristics of cats, including his sweeping assessment of black-and-white cats: "The Black-and-White Cat is affectionate and cleanly, but it is a selfish animal, and is not one for children to play with." Fortunately, this reputation of black-and-white cats did not stick.

It is black cats, though, that have the hardest time. Traditionally associated with witchcraft and evil, along with, in some countries, bad luck, they've had difficulty shaking off these stigmas. On top of this, an increasingly common reason given for not choosing a black cat nowadays is that they are difficult to take good photos or selfies with. There is also some evidence that people find it harder to read the emotions of black cats, which influences their attitude toward adopting one. One study looking specifically at adoption of black cats from shelters found that it takes around two to six days longer to rehome black cats compared to non-black ones. Even "tuxedo" and other black-and-white cats were rehomed around three days faster than completely black cats. This was the case for both adult cats and kittens.

Do these associations between coat color and personality have any factual basis? Several studies have considered this question, with mixed results. While there may be a slight tendency for some coat colors to be associated with increased aggression toward humans, on the whole, it seems that physical appearance is not a reliable predictor of personality.

Instead, there seems to be a stronger link between different breeds of cat and personality qualities. Some of the behavior characteristics of breeds are long-standing and well known. Siamese cats, for example, are notoriously vocally demanding of their owners, while Persian cats are generally described as less active or playful than other breeds. The growing popularity and development of new cat breeds has provided scientists with an opportunity to investigate the potential effects of genetics on behavior in cats. For example, through owner surveys, Milla Salonen and co-researchers discovered that breeds vary considerably in their desire for contact with people, with Korat and Devon Rex cats showing a higher tendency to seek contact with humans than British Shorthairs. Also, in some breeds, certain personality factors seem to be correlated with each other. For example, with the Ragdoll, a breed known for its laid-back personality, breeders typically select the more inactive individuals from which to breed in order to get calm kittens. However, the resulting low activity levels in this breed were found to correlate with a low tendency to seek human contact. So, in trying to produce chilled-out kittens, breeders may also unwittingly be selecting for less people-orientated progeny.

Other genetic investigations in cats have looked at specific gene locations where the presence of different variants might influence personality. For example, a variant of the oxytocin receptor (OXTR) gene has been found to contribute to a suite of "roughness" traits in cats, including irritability, dominance, forcefulness, and moodiness. The potential for future research into the existence of genetic links to different behavioral traits is enormous.

Do Birds of a Feather Flock Together or Do Opposites Attract?

There's a scene at the start of the Disney film *101 Dalmatians* in which Pongo the dalmatian watches a series of dogs and their owners walk by his window. All the owners bear an uncanny resemblance to their dogs. It's an amusing nod to the often-made joke that owners tend to look like their dogs. In a somewhat unsettling (for dog owners) study, researchers revealed that, for owners of purebred dogs, at least, the phenomenon may contain an element of truth: independent observers could match pictures of owners correctly with their dogs. Increasing attention has been given to whether owners resemble their pets in personality too. Having found some evidence of this in dogs, researchers turned to cats—do we unwittingly choose cats that match our personality?

A small but intriguing study of female cat-owning college students asked them to rate both themselves and their cats according to twelve different personality traits. Some of the women owned Siamese cats and some had mixed breeds. The owners of

Siamese cats rated both their cats and themselves similarly for the qualities of clever, emotional, and friendly, while the mixed-breed owners matched their assessments of themselves and their cats for the traits of aggressive and emotional. The results suggested one of two things: either owners see their pets as they see themselves, or owners choose pets that they think are similar to themselves.

With so many different human and cat personality types, scientists have begun to examine, just as they have for human-human relationships, how different personalities interact with one another. They have uncovered some fascinating effects. It appears that cats and their owners aren't all random mixtures of personality types; some aspects of owners' personalities are significantly related to the personalities of their cats. The personality dimension Neuroticism seems to be particularly important. In humans, Neuroticism has a strong social effect—high scores on this trait tend to result in negative outlooks that can affect those around them, be they friends, family, or pets. More specifically, cat owners who score high in this dimension typically display more needy and intense relationships with their cats, worrying about them a lot. Although their cats seem to accept the attention fairly readily, allowing their owners to pick them up and to kiss and nuzzle them, the intensity of this relationship tends to produce cats that, like their owners, are more anxious and tense.

Probing deeper into the anxious owner–cat relationship, one study found that such owners are more likely to restrict their cat's access to the outdoors and to worry that their cat has a behavioral, medical, or stress-related problem. These findings mirror the results of parent-child studies, in which anxious people (scoring high in the neuroticism dimension)

tend to display stricter, often overprotective parenting techniques as a result of worrying about their children's welfare.

Despite being more intense, cat-human relationships in which the owner scores higher on Neuroticism tend to be characterized by less rich interactions. This may be partly due to the owner initiating interactions more frequently than the cat. Dennis Turner and his co-researchers previously showed that the length of interactions between humans and cats depended on which participant started it. Interactions initiated by the cat last longer than those begun by the person. Conversely, in relationships in which the owner scores high on Conscientiousness, the cat-human interactions are more intricate and contain more behavioral elements. Talking about these results, Kurt Kotrschal and co-writers suggest that a more controlled, conscientious personality in an owner instills in the cat a sense of regularity and dependability, allowing for the development of ritualized displays between owner and cat. When owners score high on the Openness dimension, their relationship with their cat is generally different—less intense. Their cats are usually more relaxed, vocalize less, and spend less time looking at their owners, possibly indicating that they feel more secure.

The tendency for certain human character types to reflect particular cat personalities is unlikely to be a coincidence. The most likely explanation is that through their control of their cats' environment, owners unwittingly shape their cats' behavior by encouraging or discouraging certain aspects of their personalities. It may not be all one-way, however. Particularly within the more intense cat-human relationships, cats may recognize their owner's neediness—for example, with respect to eating—and use it to their advantage, perhaps becoming more finicky with food and creating a "negotiating" relationship. The resulting cycle of need and negotiation can make both owner and cat more mutually dependent and anxious. Owners with higher Openness scores may, through fussing less, encourage their cats to be more independent and able to cope with new situations. Introduction of a novel object into the room fazes these felines way less than it does anxious cats.

Given the potential combinations of different owner and cat personality traits on top of all the environmental ifs and buts and maybes along the way, finding the right cat for people can be a challenge. Part of the role of a cat rescue shelter is to try to match potential new owners with a cat whose personality complements theirs. It was my favorite part of my work in the shelter, and usually, once I had chatted with the people for a while, I could tell which of the cats in the cattery might best suit them. Most of the time, matchmaking efforts by me and my colleagues worked well. Occasionally, as the lady meeting the tortoiseshell cat Pebbles demonstrated, people would throw a curveball and we would have to rethink the match. And then there were the really special cases, the times you realized it really was worth getting up day after day, week after week, traveling on a bus in the early morning, still dark in the depths of winter, to come and do your job. Moments when you realized that, however much you know about cats and people and how they communicate, sometimes cats and people choose each other for totally unfathomable reasons.

I walked down the corridor that ran alongside the cat pens, the little girl and her parents following me. I stopped outside one where a chunky tabby cat was enthusiastically purring and rubbing enticingly against the bars, looking hopefully up at us. I turned to the little girl. "This one's name is Mimi—you might like to meet her?" The family duly took turns going into various pens to interact with first Mimi, then with a larger selection of cats I thought might be a good match for them. The little girl was gentle and thoughtful with all of the cats but stayed fairly quiet. Eventually, as we made our way down the corridor, she stopped outside a pen where there was no cat waiting hopefully at the front. "Which cat is in here?" she asked. "Oh, that's Ginny," I said. "She's been here a few weeks now but is still a bit too shy to come out and meet people. She will be hiding in her box at the back." "Can I see her?" the little girl asked. I looked doubtfully at her mom. "Well, yes, of course, but she's scared and not feeling friendly, so you will need to be very careful not to get too close to her box." With her mom's approval I tentatively opened the door and let the girl in. She made no attempt to approach Ginny's box, which was up on a shelf. Instead, she simply sat down on the floor and quietly said, "Hi, Ginny," whereupon Ginny, the black-and-white cat that so far had refused all our hard-worn attempts to coax, love, and feed her out of sadness, crept silently from her hidey-hole, jumped down, and approached the little girl. Ginny rubbed luxuriously all around her, climbed into her lap, and settled down. As I silently beckoned to all my colleagues to witness the miracle in front of me, open-mouthed with astonishment, the little girl looked up at her mom and smiled. "I think Ginny is the one for us."

CHAPTER 8

THE PLEASURE OF THEIR COMPANY

What greater gift than the love of a cat?

—Charles Dickens

A few months into my postgraduate studies, visiting or watching cats all day long, I found that when I came home at night, I missed their company. I decided I needed a cat of my own, and so, after a little searching, I adopted my very first kitten. A long-haired tabby—whom I imaginatively named Tigger—he was the progeny of a stray local tomcat and a domestic pet female cat. Why did I choose him? Like many owners say when asked, there was just something about him, even at eight weeks old. As he grew up, it became obvious Tigger was wild at heart—out for hours on end, goodness knows where, doing goodness knows what. He paid the price for this one unfortunate night when he was about nine years old, losing a leg in a road accident, but he rallied and got around admirably on three legs.

As a young cat Tigger was aloof by nature, rarely passing the time of day with any of my friends or visitors. Despite this, he was always affectionate with me—he proved to be a great companion, greeting me with a raised tail, rubbing around me, and curling up on my lap in between his forays outdoors. As he aged, he became more sociable with other people, but he still kept his special affection for me, always remaining, to some extent, at least, my cat.

Not many people like to admit whether they have a favorite pet. It feels far too mean even when, to an outside observer, it may seem quite obvious. Cats have no such qualms—some simply prefer just one person in their life and that's it, while others take to and interact with anyone who will give them attention. Either way, they make no secret of it. One of the enduring mysteries of the cat-human relationship is how or why cats are attracted to particular people—is it just down to human and cat personalities, or is there something more going on? Researchers have begun to delve deeper into the intricacies of the cat-human bond, including how it develops, what affects it, and the different types of relationships that exist.

Young domestic pet kittens like Tigger, arriving in their first permanent homes after being separated from their mother, will have been exposed to wide-ranging amounts of interaction with people by this stage in their lives. This, combined with their own personalities and the many different types of people with whom they are to share their new homes, creates an enormous amount of variability in the relationships that cats build with their owners.

With so many factors in the equation, the development of cats' interaction styles with different people is a tricky one to study. So much so that few researchers have been brave enough to try. Perhaps the most comprehensive study of its kind was conducted by Claudia Mertens and Dennis Turner in 1988. Keen to find out just how conversations with cats begin from the moment a cat meets a person for the very first time, they set up staged encounters between the two.

The study looked at how men, women, and children (aged between six and ten) interacted with cats that they had never met before. Encounters were staged in an observation room, where the participants could be discreetly recorded through a one-way window. The volunteer would take a seat in the room, into which was then released one of nineteen cats that belonged to the university colony. For the first five minutes the volunteer was asked to remain seated, looking at a book and ignoring the cat. This was followed by a second five-minute period, in which they were free to interact with the cat however they wished.

Of great interest to the researchers was how the cats behaved in those first five minutes when they received no input or cues from the person sitting in the chair. The researchers recorded information that indicated each cat's interest in making contact with a new person, specifically the timings of its first approach, first social behavior, and first actual body contact with the person. The cats' individual personalities led to the greatest variability in how they behaved. For example, some cats were bolder than others and approached more readily, some preferred physical interaction, and others liked to play. Interestingly, at this early stage before the person responded, each cat's unique behavioral style remained consistent whether the recipient was male or female, adult or child. In the second half of the experiment, however, once the person began interacting with the cat, some changes in the cats' reactions began to emerge. In particular, the frequency of cats' approaches changed according to the person's age and gender, with adults approached more frequently than children, and females more than males. These differences appeared to reflect the different interaction styles of men, women, and children.

Once they were free to move around, children stayed seated much less than the adults. Among the adults, the men stayed seated more than the women, while both the women and the girls showed a tendency to crouch down on the floor to interact with the cat. If the cat was trying to rest or retreat, a child was far more likely to follow it than an adult, boys more so than girls.

In their attempts to initiate contact with the cat, adults and children also exhibited different techniques. Almost all the adults began their interactions with a vocalization, whereas children began with a vocalization in only 38 percent of the tests. The rest of the time they either approached the cat directly or started immediately playing with or petting the cat. The quality of human vocalizations differed, too—whereas the vast majority of adults spoke in complete sentences, this was the case for only a third of the children. Another third of the children used only single words or sounds, and a third said nothing at all to the cat as they interacted with it. Adults typically continued talking to the cat throughout the interaction, while children generally stopped speaking to the cat once they were physically interacting with them.

Having looked at the very first encounters between cats and people in a staged situation, Claudia Mertens went on to study the relationships between pet cats and the families they lived with, observing encounters in their own homes. Over a whole year, she visited and studied fifty-one households containing a range of different family sizes and numbers of cats per household, totaling over five hundred hours of observations. These records of well-established cat-human relationships reinforced her earlier work on different human interaction styles. As in the previous study, children exhibited greater physical activity in their interactions with their cats, while adults were more inclined to speak to them first, especially women.

As we progress from childhood to adulthood, we seem to learn it is better to grab a cat's attention first by speaking, thereby giving the cat a chance to react, before following it up with a more physical interaction. This is strangely similar to how cats change their interaction with humans as they grow up from kittenhood. Kittens gradually learn to approach and address us with a polite meow rather than scaling our legs for attention. A built-in cat-human etiquette code, perhaps.

Mertens recorded specific elements of interactions between family members and cats, including each time the cat or human approached or distanced themselves, and how often they remained within one meter of each other. Measuring how closely human approach and retreat behavior matched that of the cats, Mertens calculated how reciprocal these elements of their interactions were. She found that this reciprocity was higher between cats and adults than between cats and older children (aged eleven to fifteen) or cats and younger children (aged six to ten).

With respect to interaction reciprocity, Dennis Turner refers to the concept of "goal meshing," first mooted for rhesus monkeys, whereby the goals of each partner align with those of the other. He and his co-researchers looked in detail at interactions they recorded between cats and their owners and analyzed how each party reacted to the other's wish to interact. They found that in some relationships the cat responded positively to an owner's desire to interact and in return the owner would also respond positively and interact when the cat wanted to. A case of "You scratch my back and I'll scratch yours," although hopefully not literally. This type of relationship resulted in a higher rate of cat–human owner interactions. In other cases both parties were less likely to cooperate in response to the other's wishing to interact. Although this latter type of arrangement inevitably results in a lower level of interaction, there is still a balance—both cat and owner are content with this level—which enables the relationship to work. This ties in with the idea that interactions between owners and their cats may, over time, become ritualized. The more owners and cats are together, the more they learn from each other and gradually develop a predictable and established routine of interaction.

To drill down into human-cat encounters even further, a study by Manuela Wedl and coworkers used a more technical method. They video-recorded encounters between cats and their owners around the cat's normal feeding time and analyzed the sequences of behaviors using a software program called Theme. This program was able to spot temporal patterns—sequences of events that followed each other in a nonrandom way—that could not have been spotted by simple human observation, and then assessed the complexity and structure of the interactions. The study found that women-cat interactions tended to contain more patterns per minute than those between cats and men, suggesting, in support of Merten's findings, that cats tended to feel more comfortable when interacting with women.

A constant feature of all these studies is that women appeared to interact in a way that the cats preferred. Of course, people don't always respond according to the age and gender stereotypes found in the studies—men, women, and children all have special relationships with their cats. As Dennis Turner points out, the cats in the first experiment didn't innately prefer one gender or age category of person, but because the people in those different categories had distinctive interaction styles, the cats responded in different ways to them. Cats prefer it when people crouch down to their level, when they vocalize before interacting, and when they don't follow them or interrupt them while resting. These are all conditions that give cats some control over the encounters.

Along the same lines, perhaps one of the simplest but most significant findings from Turner and co-researchers' human-cat interaction studies, with huge implications for people when interacting with cats, was that interactions started by the humans do not last as long as those started by the cat. In other words, cats prefer to make the first move.

Gently Does It

The studies described in this chapter provide an important reminder that humans need to think about how best to interact with cats before rushing in and expecting them to respond positively. A classic example of this was my attempt to find a temporary carer for my cat Tigger while I went away on vacation.

For many years I had covered vacation pet care simply by asking a neighbor or friend to drop in once or twice each day to feed and check on my family's animals. However, the year that Tigger became diabetic and needed twice-daily insulin shots, going on vacation suddenly became more complicated. To his credit, Tigger never made a big deal out of his shots. But it was still a bit much to ask a neighbor to take on, so as vacation time approached, I researched other options. At the vet clinic, I found an ad for an in-home pet care service that also offered medical care for an added cost. Perfect. Along came Greg for an introductory visit to meet Tigger. I say "meet," but in reality this consisted of Greg pursuing a rapidly fleeing Tigger around the house, trying to be his friend. "Oh, he's a bit reserved at first," I explained. "If you just sit quietly and let him approach you in his own time, he will soon warm up." Greg looked at me as if I were mad and continued to try to actively and noisily to coax Tigger into an interaction, completely in vain. I should have followed my instinct to look elsewhere, but remembering Greg's five-star reviews, I decided things would probably be fine and signed up for ten days of twice-daily visits for insulin shots and feeding. Day one, about twelve hours into our vacation, I received a call from a very frazzled-sounding Greg. "This cat is impossible—I can't catch him to give him his shots and he just hisses and spits at me when I corner him." Two thousand five hundred miles away, on vacation, I wondered exactly what Greg expected me to do. But again I tried to guide him as to how to win Tigger over, starting with "Don't corner him." Greg's solution, sadly, was to wear very long, protective gloves, combined with a lot of chasing around the house. After ten days, I returned to a very stressed and disgruntled Tigger.

Wracked with guilt, next time I wanted to go away I worked much harder. I searched far and wide, following recommendations and checking out endless options until I found a small bespoke cat-sitting service run by a lady called Joyce from her home. Doubtful that I could ever successfully take Tigger to someone else's home and leave him there without creating havoc, I suggested a trial run and took him there for a weekend. Joyce's house was a haven of peace and tranquility. As we entered, I set Tigger's cat carrier down and let him acclimate awhile before letting him out. Joyce completely ignored him as he sniffed around cautiously. "He'll be fine—don't worry," she said, shooing me out the door like I was an anxious mother dropping her child at preschool.

Worrying myself stupid, I resisted for twenty-four hours before calling the next evening. "Hi, how's it going?" I heard a ping in the background. "Hang on a minute," said Joyce, "the microwave just finished heating Tigger's chicken for his supper." She put the phone to one side for a moment and I heard her go off and call Tigger. "Tigger, come on in, your dinner is ready." "Is he outside?" I asked anxiously when she returned to the phone, slightly horrified as I pictured Tigger scaling the garden fence and disappearing into the sunset. "Oh, he's just having a little wander around the yard while I make his supper. Then he'll come in for his insulin shot and chicken when he's ready. We did that last night, then he curled up on his special 'Tigger' cushion on the couch next to me. We're fine—see you Monday."

Same cat, different people, yet both complete strangers to Tigger. Joyce simply let Tigger come to her on his own terms (with some delicious chicken thrown in for good measure).

Many cats don't particularly enjoy the way certain people interact with them. Some register this dislike quite obviously by moving away or exhibiting aggressive behavior, but the results of one study suggested that other cats may simply tolerate being petted by humans, rather than actively enjoying or hating being stroked. These "tolerant"

cats were found to have higher levels of glucocorticoid metabolites (GCM) in their feces than other cats, indicating that they were experiencing higher levels of stress.

With a view toward increasing how comfortable cats feel during encounters with humans, Camilla Haywood and co-researchers have developed and tested a set of "best practice" guidelines for people to consider when interacting with cats. Employing the easy-to-remember acronym CAT, these guidelines emphasize the importance of allowing the cat to have choice and control (C) over an interaction; being alert (A) to the cat's responses as the interaction proceeds; and being mindful of where they are touching (T) the cat—sticking to their preferred areas around the ears, chin, and cheeks, as described in chapter five.

The guidelines encourage people to offer their hand gently to the cat at the start of an interaction, allowing it to approach and interact if it chooses. Only if the cat makes contact—for example, by rubbing on them—should they stroke it, and if the cat moves away, they should leave it alone. While stroking, the person should monitor whether the cat continues to rub, thereby indicating that it wants to continue the interaction. They should discontinue stroking if the cat stops rubbing and moves away or shows any negative body language such as flattening its ears, fluffing its fur, or twitching its tail. Spontaneous self-grooming may also indicate the cat wants to stop.

The researchers tested out how a short training session explaining the guidelines, given to people prior to meeting a new cat in a rescue shelter, affected the subsequent behavior of the cat. They also video-recorded cat-human interactions in which no prior tuition was given. Trained observers then analyzed the encounters to give an objective assessment. Results showed that in interactions in which CAT guideline training had been given to humans, cats directed more friendly types of behaviors and less of the aggressive types toward the people and showed fewer signs of negative body language, compared with interactions in which no human training had been given. While the guidelines encourage a more restrictive approach than many people generally adopt when petting cats, results from these trials indicate that allowing cats to take the lead may ultimately allow more successful interactions and relationships with them.

The Family Cat

Why do we keep cats? Our original relationship with the wildcat undoubtedly began as a functional one—pest control. However, the predatory instinct that humans used to respect so much is now often the cat's least appreciated quality. Unlike dogs, cats have not been bred for specific tasks such as herding, guarding, or sniffing things out. Even pure breeds are bred for their looks rather than any other potentially more useful qualities.

Occasionally cat advocates, in an attempt to show that cats are as useful as dogs, have endeavored to find a helpful "role" for the cat. The prize for this must surely go to the Belgian Society for the Elevation of the Domestic Cat for their efforts in the 1870s. The society felt that the talents of cats were being underutilized and so they invented a job for them. In a bid to capitalize on the considerable skills of cats at finding their way home, thirty-seven cats were tested in a trial as potential mail deliverers. The cats were transported some distance from their hometown and released, hopefully to find their way back to their homes. One cat made it back in five hours, and all thirty-seven had wandered back within twenty-four hours. Riding on this

success, the society planned to develop their scheme and attach to each "mail cat" a letter wrapped in a waterproof covering. The cats would then be released to "deliver" their mail. Although a successful confirmation of the homing talents of cats, as a system for delivering mail, the scheme, not surprisingly, never took off.

The domestic pet cat, in contrast to most domesticated animals, seems to have escaped with mostly having just one job, and perhaps the best job of all, in fact—companionship. Most people acquire a cat these days for this reason. It was certainly the reason I got Tigger. He, like many cats, began life with a single owner—me—and as time passed, he gradually became part of a very different, more complex family unit. The concept of a "family life cycle" is often used to describe the different stages that individuals within a family go through over time. These stages include human milestones such as leaving home, finding a partner, and raising a family, during which people's roles change and develop. More recently, consideration has been given to how pets, too, often have changing roles as a family grows or shrinks.

When I adopted Tigger, for example, he reluctantly, but gracefully, shared my attention with my then-boyfriend, who later became my husband. He less gracefully accepted the presence of Charlie, a sweet-natured little female cat whom I adopted a year after Tigger, but they muddled along together until we sadly lost Charlie to illness thirteen years later. Over his nineteen years, Tigger weathered the ever-increasing chaos of family life with four children, a move from the UK to the US and then back again five years later, and no fewer than eight different houses. In that time, he went from having a lot of my attention to sharing it with more and more others. As he matured and then became an old cat, Tigger changed—the cat he became was nothing like the one he had started out as, not least because he was by then three-legged, diabetic, and in possession of only a few teeth. Like all people and animals who spend more and more years together, we learned each other's habits and quirks and formed unique relationships with him.

I felt I knew Tigger well. From that very first day bringing him home as an eight-week-old bundle of fluff, we had an understanding and mutual respect. There was most definitely affection, too, albeit on Tigger's terms. My husband and daughters had different perspectives.

So where does the average cat fit into family life? In a fascinating study, Esther Bouma and colleagues asked cat owners which one of four categories they perceived their cat to be in: "family member," "best friend," "child," or "pet." Over half the replies were "family member," a result found by other researchers too. Perhaps more surprising, though, around one-third of the owners described their cats as either a best friend or child. It seems that pet cats, rather than being simply on the periphery of people's lives, have become an important part of family life. From one perspective, this is good news for cats—they receive better care, attention, and medical help and are most certainly noticed. The danger, though, as the authors of the study point out, is that cats may, in more intense relationships with owners, become regarded as "small humans" rather than cats, and their needs and behavior may be misinterpreted.

Measuring Relationships

The type of companionship that owners get from their pets clearly varies enormously, an observation that has led scientists to devise different measurement scales that quantify the pet-owner relationship. These scales mostly consider pets in general,

Steve, my husband, knew Tigger from kittenhood:

"Tigger was an excellent cat. He was always much more yours than mine, but I did love him very much and he was certainly cool. He mellowed with age but was aloof with me till the very end."

Abbie, born when Tigger was about six:

"I was a bit scared of Tigger when I was little. I remember being afraid that he would scratch me but also very much understanding that if I didn't bother him, then he wouldn't bother me. Later in his life, I had a lot of affection toward him. I remember sitting by his basket every night and stroking him before bed. It felt like a very big change from when we were both younger."

Alice, born when Tigger was seven:

"I was scared of Tigger when I was young—he didn't like us touching him. He mellowed out as we all got older, and I was able to stroke him without being afraid. Toward the end of his life, I remember he used to get a bit confused. Sometimes we'd find him in strange places, doing strange things, but mostly he stayed napping in one spot."

Hettie, born when Tigger was eleven:

"He was grumpy and slow and could flip very easily but I was never scared. I just knew you had to pick your moments. He mellowed over time."

Olivia, born when Tigger was sixteen:

"I just remember watching him in the garden, seeing off the fox."

rather than focusing on the cat-owner relationship. However, one designed specifically with cats in mind is the Cat Owner Relationship Scale, devised by Tiffani Howell and coworkers, based on the similar Dog Owner Relationship Scale. Starting with a questionnaire given to owners regarding different aspects of their relationship with their cat, the researchers use the answers to produce a score on three separate subscales. Thus the "Pet-Owner Interactions" subscale is based on questions such as "How often do you play games with your cat?" while the "Perceived Emotional Closeness" subscale includes owner ratings on "My cat gives me a reason to get up in the morning," as well as other topics. On the "Perceived Costs" subscale the owner records, among other statements, how strongly they feel regarding "My cat makes too much mess."

Including negative as well as positive components gives an overall balance indicating how rewarding the relationship is between owner and cat. The authors based this on social exchange theory, which dictates that any relationship will be maintained only if the perceived positive aspects (like emotional closeness) outweigh the negatives (like the costs of the relationship) or if negative and positive components are equal. Sadly, it is usually the cats in the net-negative relationships that end up in rescue shelters looking for new homes. Unfortunately, it is hard to know how positively cats rate their relationships with their owners. Cats that are unhappy may demonstrate their feelings, if they are able, by simply upping and leaving, potentially finding another house down the street to move into instead. Others may not have this luxury and must make the best of their situation.

The benefits enjoyed by owners when the relationship with their pet is positive are often referred to as the "pet effect." Measuring this effect has proven something of a conundrum for scientists, particularly with respect to how living with pets affects us emotionally. Again, questionnaires are generally used, asking owners how they feel their pet helps them. People often report reduced loneliness, improved self-esteem, and lower levels of depression. However, these questionnaire answers are usually given by owners retrospectively—so they are remembering how their cat made them feel rather than recording their thoughts contemporaneously as they interact with their cats. Some owners find it more difficult to recall details than others, or potentially have slightly rose-tinted memories, so although useful, these questionnaires can produce inconsistent results, which are also hard to quantify.

One set of researchers found a new approach to exploring the pet effect: studying it in real time. Participants were recruited who owned a dog, a cat, or both. Using what is known as the experience sampling method, they were asked at ten random points in the day, for five days in a row, to record their current activity and feelings, whether their pet was present, and whether they were interacting with it at that moment. They were given eleven different adjectives to choose from, both positive and negative, to describe their mood. The results were intriguing and suggested that the pet effect might be more subtly complex than previously thought. The mere presence of a pet, without even interacting with it, resulted in less negative feelings for the owner but didn't produce any more positive ones. Interaction with the pet, however, both lowered negative feelings and increased the use of positive adjectives. Although there is a slight possibility that owners interact more with their pets when they are already feeling more positive, the more likely explanation for this result is that interaction with companion animals promotes emotional well-being in owners. The researchers did not separately analyze the results for cat and dog owners—it would be interesting to know whether the same effect would be recorded for just owners of cats.

The Cat-Human Bond

As the number of companion animals has increased worldwide, scientists have become interested in the development of bonds between owners and their pets—what affects levels of attachment, and how does attachment impact the relationship between owner and pet? Looking more specifically at cats, research has found that the personality of an owner helps predict how attached they are to their cat. For example, the Conscientiousness dimension, from the Big Five discussed in chapter seven, seems to be particularly important for the human-cat relationship—people who score higher in that dimension consistently display higher attachment to their cats. Owners scoring high on the Neuroticism dimension, which includes the trait of anxiety, also tend to show high levels of attachment, as perhaps they seek extra emotional support from their pet.

People whose cats make frequent physical contact tend to be more attached to their cats than owners whose cats avoid such contact—a pleasing reminder of the power of touch from chapter five. In addition, more-attached owners have been shown to endow their cats with more human qualities—in other words, to anthropomorphize more than less-attached owners.

We all know in our own minds how we feel about our cats. One of the most elusive questions in our lives with them, though, is whether they actually care about us. Dogs are far more straightforward in this department. They, on the whole, wear their hearts on their sleeves: they follow us around, watch us relentlessly, and, sadly, often feel distressed when separated from us. With their reputation for independence and aloofness, it is often assumed that pet cats have no actual attachment to their people. Various scientists have set about testing this.

Several studies have designed experiments to show whether cats exhibit classic attachment to their owners. "Attachment theory" is a phrase coined during research originally designed to examine the psychological relationship between young children and their caregivers. In developmental psychologist Mary Ainsworth's Strange Situation Test (SST), a child was put in an unfamiliar room containing toys, along with their mother or other caregiver. The child, who remained in the room throughout the experiment, was then tested for their reactions to the temporary absence and then return of their caregiver, plus the introduction of a stranger into the situation. The idea was to see whether the child used their caregiver as a secure base when faced with an unfamiliar situation, and what type of attachment they displayed.

Various researchers studying the relationship between people and their dogs have adapted the SST by substituting a dog in the place of a child and an owner for a caregiver. Some studies have indicated there may be several different types of attachment of dogs to owners, similar to those shown by children and their caregivers. However, it has also been observed that the varied behavior of the owners in such studies may affect the resulting behavior of their dogs, so the tests may not simply measure dog attachment. It may be better to analyze the behavior of both owner and dog to get a more accurate representation of the relationship.

Inevitably, cat scientists have tried using the same technique to explore the relationship of cats with their owners. Three different attachment studies have used modified cat forms of the SST. Two of these found that the test cats showed a form of secure attachment to their owner, while the other suggested that the cats did not regard their owners in this way. This discrepancy in results may be due, at least in part, to the slightly different designs of the three SST experiments. It is also possible, and perhaps more likely, that cats don't necessarily regard or interact with people in the same way as human children or dogs do, making such tests less relevant for them.

To try to move away from simple attachment scores for both humans and cats and find a better overall way of describing cat-human relationships, Mauro Ines and co-researchers carried out a complex questionnaire-based study that drilled down into the various components of such relationships. Based on responses from 3,994 people, they identified five different types of relationships between cats and owners, driven mainly by four factors: the owner's emotional investment in the cat; the cat's acceptance of others; the cat's need for owner proximity; and the cat's aloofness.

In two of these relationship types, the owner shows relatively low investment in the cat emotionally. One of these, described as a "remote" relationship, is also characterized by the cat's low sociability—they display little need to be near their owner or humans in general. In the "casual" relationship, cats are more sociable than in the remote category, but they tend to associate with people in general, not favoring their owner. The authors suggest these casual relationships may be typified by cats that come from busy households and have outdoor access, and that tend to visit neighboring households.

In the "open" style of relationship, owners display a moderate amount of emotional investment in their fairly independent cat, which typically enjoys the company of people but does not seem to specifically seek out their owner. Unlike those in the other groups, these are the cats most likely to be described as "aloof" by their owners, resembling Kipling's famous "cat that walked by himself." The result is a weak but evenly matched cat-human bond.

The other two relationships identified by Ines and colleagues include owners who invest a lot in their cats emotionally. In one, the "friendship" relationship, the cat is sociable but flexible in who they will interact with. They enjoy being with the owner but are less emotionally tied to them compared with the more intense "codependent" relationship. In the latter, both cat and owner display a strong emotional bond with each other, spending large amounts of time together. The cat may be so attached that they won't interact with other people. These types of cats may show separation-related problems, such as destructive behavior or inappropriate urination when their owner is not available for interaction. This sort of relationship is more likely to occur in one-person households and often with indoor-only cats. Traditionally associated with dogs who become dependent on owners being present, separation anxiety is an increasing concern for cats that rely highly on their owners for company and stimulation.

How Do You Feel About That?

Tigger was most definitely not the type of cat who needed the constant attention of either me or the rest of the family, but I often wondered how he felt during those ten days I'd left him in the care of Greg, and how that compared with his later stay with Joyce. Was he scared, angry, sad, or stressed around Greg? At

Joyce's did he feel happiness? Do cats actually experience the same emotions in life as we do? To suggest such a thing would, at various points in history, have been met with accusations of the dreaded anthropomorphism from philosophers and scientists alike. A change in attitude toward animal emotions has gradually emerged, though, as scientists discovered that humans and non-human animals share some of the more basic emotions such as fear, which is processed by the brain's limbic system, and the physical responses that occur as a result.

For example, a human in a fearful situation—such as suddenly finding themselves in the path of an oncoming vehicle—will experience a surge of adrenaline leading to a fight-or-flight response. Their heart races, their breathing accelerates, their pupils dilate, and they may get goose bumps as the fine hairs on their body rise. A frightened cat, perhaps encountering a large barking dog, or even a rapidly advancing Greg, will undergo similar physiological changes. Their heart and respiration rate will increase, and their pupils dilate. Along with these, their fur stands up on end (piloerection) in what must be said is a far more impressive display than in humans. This "Halloween cat" effect makes them appear to the source of the fear to be much larger than they actually are.

These are similar physical responses to those of humans, but it is impossible to really know whether cats experience emotions in the same way we do. Nevertheless people, and particularly pet owners, often assume that animals feel things as we do. We attribute to our pets the basic emotions of fear, anger, joy, surprise, disgust, and sadness. Complex emotions, such as jealousy, shame, disappointment, and compassion, are less commonly used by owners when describing their pets, although dogs tend to be credited with more of them than cats, possibly due to the more developed sociality of dogs. That said, I found when working as a behavior counselor that pet cats exhibiting "problem" behavior would occasionally be credited with complex motivations such as jealousy or spite—as illustrated by Mrs. Jones, when she caught Cecil spraying on her boots back in chapter two.

One emotion that most scientists agree domestic cats, particularly pet cats, appear to experience more and more is stress. Despite being impressively adaptable and managing to live in almost any environment, many cats find modern-day life with humans and other cats enormously stressful. A little stress, the sort that brings the occasional rush of adrenaline to the system as described in the preceding scary dog scenario, is normal and manageable for a cat. However, sometimes cats find themselves in ongoing stressful environments, and this chronic stress can lead to physical and behavioral problems. Cats may become more withdrawn or jumpy; develop uncharacteristic new behaviors (like Cecil and his boot-spraying behavior); or, in more chronic cases, start to overgroom and pull their fur out or develop stress-related illnesses.

The cause of the stress is often simply living alongside other cats, something that frequently comes as a surprise to cat owners given the cat's ability to live in colonies when necessary. The difference is that colony cats, such as Ginger and Sid in the group I studied on the farm and Tabitha and Betty in the hospital group, can generally pick and choose with whom they interact. If a certain cat is intimidating or unfriendly, they avoid them. On top of this, cats in colonies are often related to some extent, as the social system of such groups is based around generations of females from the same family. Pet cats, on the other hand, are frequently faced with an unrelated feline "friend" coming to live with them, often after they have lived a considerable period of time as the only cat in their home. Slow and careful introduction of new cats is very important, but even then, they may not necessarily get along. Adopting two siblings from a young age is usually more successful, although they, too, may be incompatible when they reach adulthood (just like Bootsy and Smudge in chapter five). In one survey of owners with more than one cat, over 70 percent reported that there were signs of conflict right from the moment of introduction. Although in many cases this eases over time, the underlying tension is often still present, with staring reported as occurring daily in 44 percent of the multi-cat households and hissing daily in 18 percent of them.

In such a domestic situation, with two or more cats living in one home and sharing the same resources, opportunities for avoiding one another may be limited, particularly for indoor-only cats. They may have to share litter boxes, feeding locations, and resting spots, and confident cats sometimes very subtly bully other cats in a household by staring at them, or by blocking doorways, cat flaps, or litter trays. Meanwhile, while some pet cats may enjoy the freedom of going outside, this may be tempered by their stress at potentially encountering unfriendly neighborhood cats. Cats from separate homes don't generally regard one another as being from the same social group, which makes hostilities more likely. In densely populated residential areas, the amount of territory available for any one cat is small, and competition for space can be fierce.

Solving this problem involves finding the root of the stress. If other household cats are the source, then increasing resources such as litter boxes, feeding locations, and hiding and resting spots may reduce conflict between resident cats and enable them to share the space more easily. When outside cats are the source of stress, making sure the resident cat feels secure inside the house is important by

preventing other cats from coming through flaps or possibly covering up windows where outside cats can constantly be seen.

Reading Each Other's Minds (and Faces)

Recognizing the emotions of others is an important ability for many animals—not just from a social perspective but also for survival. Spotting that those around are in a heightened emotional state, particularly fear, tends to produce a similar reaction in the observer—so-called emotional contagion. Humans do this, too—we look to each other to see what the other person is feeling or how they are reacting to something, particularly in child-adult relationships, focusing particularly on facial expressions to give us clues. Scientists have developed a more objective method of describing human facial movements in terms of underlying muscle movements. These movements are known as Action Units (AU), and form part of the Facial Action Coding System (FACS), as discussed in chapter four. On a day-to-day basis, though, we still look at each other and subjectively try to read each other's minds from our expressions.

We look at the faces of our companion animals to try to read their expressions, too, particularly dogs. Alfie, our old golden retriever, had an amazingly expressive face. He could cock one eyebrow completely independently of the other and change from looking what we affectionately thought of as "surprised" to "quizzical" to "hangdog" in a matter of seconds. "Aw, look at his sad face," people would say.

Studies of the facial anatomy of dogs, leading to the development of the Dog Facial Action Coding System (DogFACS), revealed that our canine companions have something of a secret weapon in their communication repertoire. It takes the form of one small but surprisingly strong muscle above each eye known as the levator anguli oculi medialis, which on its contraction enables them to raise their inner eyebrow. This movement, coded as AU 101 in the DogFACS, not only makes dogs' eyes seem larger, giving them a more infantile look (so-called puppy dog eyes), but also echoes the expression humans make when sad. The eyebrow raise is a powerful elicitor of emotion in humans—researchers found that dogs in a rescue shelter who exhibited more of this expression were rehomed faster.

Our cat Smudge, who had no time or patience for Alfie, or indeed any dog, would have raised her eyebrows skyward at hearing such anthropomorphic sentiment over Alfie's "sad" looks. Except, of course, she couldn't. Cats, lacking the well-developed levator anguli oculi medialis muscle of dogs, simply do not have "puppy dog eyes." In the book *Anatomy of the Cat*, by Jacob Reighard and H. S. Jennings, the nearest equivalent muscle in cats is named the equally tongue-tying corrugator supercilii medialis. Described as "a thin sheet of scattered fibres," this is basically a weak band of muscle draped across the ridge of the eyes. It helps open the upper eyelid but does not have any impact on eyebrow movement. No wonder cats struggle to look even remotely surprised or quizzical, let alone hangdog.

Although cats lack any eyebrow-raising ability, CatFACS revealed that they have a surprisingly wide range of facial movements involving other sets of muscles, such as those controlling ear movements described in chapter four and whiskers in chapter five. Developers of CatFACS recorded the facial movements of cats in a rescue shelter and compared them with how fast the cats were adopted. Interestingly, they found no correlations at all. Unlike the persuasive eyebrow raising of dogs, facial expressions of cats seemed to have no effect on prospective adopters. It is possible that with cats, people tend to look for other, more obvious behavioral signals rather than subtle facial changes. Indeed, the only behavior that appeared to affect the rate of adoption of cats in the CatFACS study was object rubbing against the cage door, a particularly visible behavior that is familiar to many people.

Researchers Lauren Dawson and colleagues looked more specifically at how successful humans are at deciphering the facial expressions of unfamiliar cats and determining how they are feeling. They recruited over six thousand volunteers via an online survey and, using one of the most lucrative modern sources of cat videos, YouTube, tested them for their ability to distinguish between cats showing positive and negative facial expressions. People varied widely in their ability to complete this task, with women, younger people, and those possessing professional experience with cats generally achieving higher scores. However, most people found it difficult, scoring below chance. The study also found an added subtle but surprising effect regarding attachment to cats in both men and women. The higher a person's score of attachment to cats, the better they were at recognizing positive cat expressions but the *worse* they were at deciphering negative expressions. The authors suggest that more highly attached cat owners might be used to focusing more closely on signs that their own cats are happy, in contrast to owners who aren't so close to their cats and who may be more used to their own cats showing negative expressions.

Looking at things from the cat's perspective, researchers Moriah Galvan and Jennifer Vonk set out to discover how good cats are at reading human emotions by testing whether they reacted differently to human facial expressions of happiness and anger. Cats' reactions were recorded toward both their owner and a stranger, both of whom showed facial expressions indicative of either anger (frowning, clenched fists, puckered mouth) or happiness (relaxed hands and face, smiling). The researchers found that the cats took the same amount of time to approach their owner whether they looked angry or happy. After approaching, however, cats spent more time in contact with and performed more positive behaviors toward their owner in the happy compared with the angry condition. When interacting with the stranger, they showed no differentiation in behavior according to whether the stranger was "happy" or "angry." In other words, the cats seemed slightly more in tune with the expressions of their owners compared with strangers. These results don't necessarily mean that the cats understood that the owner was sad or angry—more that they had learned that different consequences usually result from the appearance of happy or angry cues. This may explain why they reacted differently with their familiar owner, with whom they may have experienced such cues in the past, compared with the stranger, whose face they had never looked at before.

A later study by Angelo Quaranta and colleagues tested cats' ability to match emotional sounds with the correct visual picture. Seated on their owner's lap, each cat was shown two photographs side by side of an unfamiliar person on a screen, one displaying anger and one happiness. Researchers simultaneously played a sound recording of either a voice laughing, a voice growling, or a third alternative: a control sound (known as a Brownian sound). They then recorded how long cats looked at each of the photographs, discovering that they looked longer at the photos that matched the vocalization being played. This suggested that the cats had an expectation of what that expression should sound like. In addition, they evidently realized that the outcome of an angry picture/growling sound combination was likely to be negative as, in addition to looking longer at the picture, they also displayed increased levels of stress behaviors such as tail tucked under and down and flattened ears.

Reading each other's emotions is a tricky challenge for two species who naturally communicate in such different ways. Cats have an excuse for finding it difficult—their solitary, poker-faced ancestors rarely had to interact with one another and so relied largely on scent for communication. And yet, despite our instinctive fascination with faces, expressions, and emotions, it seems that cats, to some extent, may be better at reading us than vice versa. Hopefully resources such as CatFACS will become more widespread, and we will learn how to better interpret cats' facial movements. In the meantime, thankfully, cats have devised ways to show us how they are feeling using their other visual, vocal, tactile, and scent signals.

Many owners watching their cats as they approach wonder: Do they seek us out simply for the pleasure of our company? One study put this to the test, giving cats a choice between social interaction with a person, some tasty food, a tantalizing toy, and an interesting scent on a cloth. There was huge variation in the cats' individual preferences, but reassuringly 50 percent of the cats chose social interaction over the other categories. Thirty-seven percent of them preferred food to other options, 11 percent chose a toy, and only 2 percent selected the scented cloth. Cat domestication has, it seems, progressed enough to make at least some of our pet

cats look to us for companionship, as we do with them. However, as my wild-at-heart Tigger taught me only too well, the opportunistic wildcat of the Fertile Crescent is really only a whisker away.

It was a summer evening, and I was standing at the kitchen counter preparing the evening meal, with the back door open onto the garden. Tigger was still a young cat then and I was suddenly aware of him there in the kitchen with me. I glanced down as he rubbed luxuriously around my legs, purring gloriously. "Hi there, Tigs," I said, smiling, flattered at how pleased he was to see me. He continued with his elaborate rubbing ritual back and forth as I chatted away to him. This went on for some time while I prepared chicken for a casserole, and then I turned around to wash my hands at the sink. Quick as a flash, Tigger was up on the counter, chicken in his mouth and back down, out through the open door and gone, along with my chicken. In hindsight, and with just a tinge of disappointment, I realize that Tigger would probably have been in the "prefers food" category of cats according to the study described. I like to think I came a close second, though.

EPILOGUE

THE ADAPTABLE CAT

The cat alone has after all solved the biggest problem posed to any animal: how to live amicably with man, and yet be absolutely free of him!

—Katharine Simms, *They Walked Beside Me*

I stood in the kitchen watching Smudge peer out through the cat flap. Since kittenhood we had kept her and her sister, Bootsy, in at night for safety, setting their cat flap in the early evening to "in only" so that they could get in, but not back out, later in the night. What blessed lives they have, I thought, with everyone so concerned for their welfare. Smudge had made the mistake of coming in earlier than usual this evening. Now, presumably hoping for a late-night stroll around the neighborhood, she pushed her nose repeatedly and unsuccessfully against the flap. She looked at me hopefully. Unrelenting, I kept it locked.

A week or so later I came downstairs to the kitchen to find Bootsy waiting for her breakfast, but no Smudge. Puzzled, I checked the cat flap and saw that the slider that denoted the settings was on "in and out." I distinctly remembered setting it to "in only" the night before. "Who came down early today and opened the flap?" I demanded of my family at breakfast. Blank looks all around.

The same thing happened the next night, and the next. Mystified as to how this great escape was happening, I found myself staking out the cat flap at every opportunity. It was late one evening, several days later, when I finally witnessed the crime in action. Unaware that I was spying on her, Smudge approached the already locked flap and tested it with her nose to see if it was open. Undeterred by its resistance, she then sat in front of it and gently nudged the slider along a little with her front paw. Then a little more, pushing the flap with her nose as she did so to see if it would release. Eventually she slid it far enough and she was out, off on her nighttime wanderings.

The manufacturers of the cat flap were amazingly responsive. "We're working on a new design," they said, "to stop cats like yours opening the flap." I obviously wasn't alone with my Houdini cat. A few months later, they sent me a new version of the flap. Instead of the slider, this one had a dial that you turned to the four different settings. A dial that required an opposable thumb to turn. "Sorry, Smudgey," I said to her as she watched me replace the old flap with this new model. "You'll never be able to open this one."

As I pondered Smudge and her surprisingly clever defeat of the original cat flap slider, I thought about the many different challenges facing domestic cats, alongside the greatest one of all, communication.

This book has shown just how impressively cats in all walks of life have learned how to communicate both with each other and with humans to succeed in our crowded and very social, anthropocentric world. From a solitary background, they have adapted to living alongside other cats, finding new signals and ways to avoid conflict in order to share resources. Even more impressively, they have tapped into our very different human language and adapted their own limited bank of communicative behaviors to fit in with ours, to grab our attention and try to tell us what they want. They understand us far better than most people give them credit for, and far better than we understand them. Has cat-human communication evolved as far as it can? It seems unlikely, given the talent that cats have for fitting into our world.

Take the average pet cat contending with the daily task of communicating with people and other cats (either indoors or out, or both). On top of this, they have to cope with a whole host of environmental challenges as a result of living in human homes. With their very different sensory world, they navigate the sights (strangely shaped), sounds (mostly noisy), and smells (very strong) of the numerous household features we take for granted such as doors, windows, taps, toilets, televisions, washing machines, and dishwashers. Even some of the human-made conveniences we provide specifically for our cats can take some getting used to: litter boxes; cat beds of all shapes, sizes, and materials; and, for those allowed outdoors, perhaps a cat flap too.

Cat flaps are an interesting modern conundrum for cats—a concept that would be totally alien to the cat's solitary wildcat ancestor. They actually started off as cat holes, cut into barn walls or doors to allow cats in to dispatch the rodents causing havoc in the stores of grain inside. A hole in a wall is easy for a cat to navigate, but somewhere along the way, someone decided to add a flap to it. Nowadays, cat flaps come in all varieties—from simple in-out ones to multifunction ones that can keep selected cats in or unwanted cats out of a home. There are even flaps that can read a cat's microchip to allow its exclusive entry.

Some pet cats learn to use a cat flap fairly quickly, following a few human-assisted postings in and out with the flap propped open, accompanied by well-timed treats. Others never quite take to it and continue to sit resolutely outside the back door waiting for it to be opened, unimpressed by the modern-day requirement to squash their face up against a plastic surface to get in or out. Then there are the cats that appear to embark on a lifelong mission to defeat the locked cat flap. Cats like Smudge.

About six months after I had installed the new dial-operated cat flap, I noticed telltale early-morning absences of Smudge once more. I checked the cat flap over and over, but all seemed fine. Again, I found myself skulking around the cat flap late at night to see what was going on. I didn't have to wait long. Watching from a kitchen chair a couple of nights later, I saw Smudge approach the flap, set to "in only" for the night. She nudged the flap to see if it would open. Finding it wouldn't budge, Smudge deftly hooked one of her claws under the bottom edge of the plastic door and pulled it *inward*, toward her, tucked her nose in the gap, and wriggled herself under and out, liberating herself once more. Ingenious. Who needs opposable thumbs anyway?

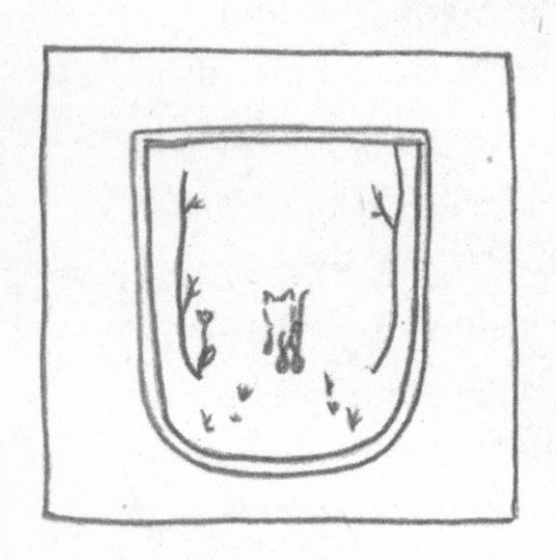

ACKNOWLEDGMENTS

Firstly, I would like to thank unreservedly the numerous scientists whose dedicated and scrupulous studies have resulted in so many intriguing and enlightening discoveries, scattered throughout this book, of how cats tick. It is said "Time spent with cats is never wasted," however, knowing only too well the challenges and the joys of studying and working with cats, I take my hat off to you all. I hope I have done your work justice. If I have made any mistakes, I apologize.

My wonderful literary agent, Alice Martell: thank you for taking a leap of faith and loving the idea of my book as much as I did, for your constant encouragement, and for answering my endless questions. Likewise to Stephen Morrow and Grace Layer at Dutton. Thank you both for the reading and rereading of drafts and more drafts, and for your inspired yet patient, gentle, and kind suggestions of how to make my chapters better. And to everyone else at Dutton involved with making this book, including Diamond Bridges, Alice Dalrymple, Isabel DaSilva, Tiffany Estreicher, Jillian Fata, Sabila Khan, Vi-An Nguyen, Hannah Poole, Nancy Resnick, Susan Schwartz, and Kym Surridge.

Thank you to all the people I have had the privilege of working with, right from the early days at the Anthrozoology Institute at Southampton University through to now. The list is far too long to write here (and I would only end up missing someone), but you all know who you are. A particular thank-you to John Bradshaw, who started my cat career rolling as my PhD supervisor and boss at the Anthrozoology Institute. To everyone who works in animal rescue—thank you for the work you do, day in day out, to improve the lives of countless animals.

My own pets deserve a mention—my cats over the years: Tigger, Charlie, Bootsy, and Smudge. Thank you for sharing your lives with me, and for sitting on my lap, desk, papers, and keyboard as I tried to write. Any typos were down to you! And my dogs: Alfie and now Reggie—thank you for dragging me out of the house to walk and clear my mind every single day, come rain or shine.

To my dear mum and dad who I wish were still around to read this book—thank you. And to my family, what can I say—I simply couldn't have written this without you all. My wonderful daughters Abbie, Alice, Hettie, and Olivia: thank you for always listening—for the chats, words, suggestions, draft readings, cups of tea, and endless encouragement on the days where writing eluded me. Hettie, thank you for magically turning my garbled thoughts and scribbles into such beautiful line drawings. And, finally, to my husband, Steve—from those early days keeping me company while I fed the ferals through the many years of children, pets, and rescued cats and kittens, you've been by my side cheering me on. Thank you for everything.

NOTES

INTRODUCTION

2 **In the USA alone:** "Pets by the Numbers," Humane Society of the United States, https://humanepro.org/page/pets-by-the-numbers, accessed July 12, 2022.

CHAPTER 1: WILDCATS AND WITCHES

12 **In 1868, Charles Darwin noted:** Charles Darwin, *The Variation of Plants and Animals Under Domestication* (London: John Murray, 1868).

12 **famous ongoing domestication study:** Dmitri Belyaev, "Destabilizing Selection as a Factor in Domestication," *Journal of Heredity* 70, no. 5 (1979): 301–8; Lyudmila Trut, Irina Oskina, and Anastasiya Kharlamova, "Animal Evolution During Domestication: The Domesticated Fox as a Model," *BioEssays* 31, no. 3 (2009): 349–60.

13 **studies have come under deeper scrutiny:** Kathryn A. Lord et al., "The History of Farm Foxes Undermines the Animal Domestication Syndrome," *Trends in Ecology and Evolution* 35, no. 2 (2020): 125–36.

13 **Some of these urban foxes:** Kevin J. Parsons et al., "Skull Morphology Diverges Between Urban and Rural Populations of Red Foxes Mirroring Patterns of Domestication and Macroevolution," *Proceedings of the Royal Society B: Biological Sciences* 287, no. 1928 (2020): 20200763.

14 **4 percent of owners in the US:** "Pets by the Numbers," Humane Society of the United States, https://humanepro.org/page/pets-by-the-numbers, accessed July 12, 2022.

14 **8 percent in the UK:** *Cats Report UK 2021,* Cats Protection, https://www.cats.org.uk/media/10005/cats-2021-full-report.pdf.

15 **Kittens must first meet humans:** The socialization period of cats was discovered in a series of experiments by Eileen Karsh in the early 1980s, described in Eileen B. Karsh and Dennis C. Turner, "The Human-Cat Relationship," in *The Domestic Cat: The Biology of Its Behaviour,* ed.

Dennis C. Turner and Patrick Bateson (Cambridge, UK: Cambridge University Press, 1988), 159–77.

17 **study of the DNA of the entire cat family:** Stephen J. O'Brien and Warren E. Johnson, "The Evolution of Cats," *Scientific American*, July 1, 2007.

17 **the catlike Pseudaelurus:** More about Pseudaelurus and its predecessors can be found in Sarah Brown, *The Cat: A Natural and Cultural History* (Princeton, NJ: Princeton University Press, 2020), 14–7.

17 **A groundbreaking study by Carlos Driscoll:** Carlos A. Driscoll et al., "The Near Eastern Origin of Cat Domestication," *Science* 317, no. 5837 (2007): 519–23.

17 **wildcat), *Felis lybica lybica*:** *Felis lybica lybica* was previously referred to as *Felis [silvestris] lybica*—this changed following a revision of the taxonomy of the Felidae family in 2017. Andrew C. Kitchener et al., *A Revised Taxonomy of the Felidae. The Final Report of the Cat Classification Task Force of the IUCN/SSC Cat Specialist Group, CATnews* Special Issue 11 (Winter 2017).

18 **Eric Faure and Andrew Kitchener estimated:** Eric Faure and Andrew C. Kitchener, "An Archaeological and Historical Review of the Relationships Between Felids and People," *Anthrozoös* 22, no. 3 (2009): 221–38.

18 **tamable and untamable species:** Charlotte Cameron-Beaumont, Sarah E. Lowe, and John W. S. Bradshaw, "Evidence Suggesting Preadaptation to Domestication Throughout the Small Felidae," *Biological Journal of the Linnean Society* 75, no. 3 (2002): 361–6.

19 **son and heir of Akbar the Great:** Charles A. W. Guggisberg, "Cheetah, Hunting Leopard (*Acinonyx jubatus*)," in *Wild Cats of the World* (London: David & Charles, 1975), 266–89.

20 **taming Scottish wildcat kittens:** Frances Pitt, *The Romance of Nature: Wild Life of the British Isles in Picture and Story*, vol. 2 (London: Country Life Press, 1936).

21 **Subsequent studies, delving into genetic:** Claudio Ottoni et al., "The Palaeogenetics of Cat Dispersal in the Ancient World," *Nature, Ecology and Evolution* 1 (2017): 0139; Claudio Ottoni and Wim Van Neer, "The Dispersal of the Domestic Cat: Paleogenetic and Zooarcheological Evidence," *Near Eastern Archaeology* 83, no. 1 (2020): 38–45.

21 **These "barnyard" animals all had:** Carlos A. Driscoll, David W. Macdonald, and Stephen J. O'Brien, "From Wild Animals to Domestic Pets, an Evolutionary View of Domestication," *PNAS* 106, suppl. 1 (2009): 9971–8.

23 **some areas of continental Europe:** Mateusz Baca et al., "Human-Mediated Dispersal of Cats in the Neolithic Central Europe," *Heredity* 121, no. 6 (2018): 557–63. Also Ottoni et al., "The Palaeogenetics of Cat Dispersal in the Ancient World."

24 **shaving off their eyebrows:** Herodotus wrote about this in his *Histories*; see Donald W. Engels, *Classical Cats: The Rise and Fall of the Sacred Cat* (London: Routledge, 1999).

24 **"an accident of history":** Faure and Kitchener, "An Archaeological and Historical Review."

26 **leopard cats *(Prionailurus bengalensis)*:** Jean-Denis Vigne et al., "Earliest 'Domestic' Cats in China Identified as Leopard Cat (*Prionailurus bengalensis*)," *PloS One* 11, no. 1 (2016): e0147295.

26 **may have gradually nudged:** Ottoni and Van Neer, "The Dispersal of the Domestic Cat."

30 **following a model proposed:** Raymond Coppinger and Lorna Coppinger, *Dogs: A Startling New Understanding of Canine Origins, Behavior and Evolution* (New York: Scribner, 2001).

30 **intriguingly, for us humans:** Brian Hare, "Survival of the Friendliest: *Homo sapiens* Evolved via Selection for Prosociality," *Annual Review of Psychology* 68, no. 1 (2017): 155–86.

30 **As Driscoll and coauthors observed:** Driscoll, Macdonald, and O'Brien, "From Wild Animals to Domestic Pets."

30 **often referred to as "facultatively social":** Kristyn R. Vitale, "The Social Lives of Free-Ranging Cats," *Animals* 12, no. 1 (2022): 126, https://doi.org/10.3390/ani12010126.

31 **Studies of these have shown:** David W. Macdonald et al., "Social Dynamics, Nursing Coalitions and Infanticide Among Farm Cats, *Felis catus*," *Advances in Ethology* (supplement to *Ethology*) 28 (1987): 1–64.

31 **Even in neutered colonies:** Sarah Louise Brown (unpublished data), "The Social Behaviour of Neutered Domestic Cats (*Felis catus*)" (PhD diss., University of Southampton, 1993).

CHAPTER 2: MAKING SCENTS

34 **the same level of nutritional:** Robyn Hudson et al., "Nipple Preference and Contests in Suckling Kittens of the Domestic Cat Are Unrelated to Presumed Nipple Quality," *Developmental Psychobiology* 51, no. 4 (2009): 322–32, https://doi.org/10.1002/dev.20371.

34 **Puppies show no tendency:** Lourdes Arteaga et al., "The Pattern of Nipple Use Before Weaning Among Littermates of the Domestic Dog," *Ethology* 119, no. 1 (2013): 12–9.

34 **kittens did not instinctively find:** Gina Raihani et al., "Olfactory Guidance of Nipple Attachment and Suckling in Kittens of the Domestic Cat: Inborn and Learned Responses," *Developmental Psychobiology* 51, no. 8 (2009): 662–71.

35 **many sources of smell:** Nicolas Mermet et al., "Odor-Guided Social Behaviour in Newborn and Young Cats: An Analytical Survey," *Chemoecology* 17 (2007): 187–99.

35 **"Are You My Mummy?":** Péter Szenczi et al., "Are You My Mummy? Long-Term Olfactory Memory of Mother's Body Odour by Offspring in the Domestic Cat," *Animal Cognition* 25 (2022): 21–6, https://doi.org/10.1007/s10071-021-01537-w.

36 **Mother cats can tell:** Oxána Bánszegi et al., "Can but Don't: Olfactory Discrimination Between Own and Alien Offspring in the Domestic Cat," *Animal Cognition* 20 (2017): 795–804, https://doi.org/10.1007/s10071-017-1100-z.

36 **kittens within a litter develop:** Elisa Jacinto et al., "Olfactory Discrimination Between Litter Mates by Mothers and Alien Adult Cats: Lump or Split?" *Animal Cognition* 22 (2019): 61–9, https://doi.org/10.1007/s10071-018-1221-z.

37 **sniffs comprising 30 percent:** Kristyn R. Vitale Shreve and Monique A. R. Udell, "Stress, Security, and Scent: The Influence of Chemical Signals on the Social Lives of Domestic Cats and Implications for Applied Settings," *Applied Animal Behaviour Science* 187 (2017): 69–76.

43 **Early studies by Warner Passanisi:** Warner Passanisi and David Macdonald, "Group Discrimination on the Basis of Urine in a Farm Cat Colony," in *Chemical Signals in Vertebrates 5*, ed. David Macdonald, Dietland Müller-Schwarze, and S. E. Natynczuk (Oxford, UK: Oxford University Press, 1990), 336–45.

44 **A detailed investigation:** Chiharu Suzuki et al., "GC × GC-MS-Based Volatile Profiling of Male Domestic Cat Urine and the Olfactory Abilities of Cats to Discriminate Temporal Changes and Individual Differences in Urine," *Journal of Chemical Ecology* 45 (2019): 579–87, https://doi.org/10.1007/s10886-019-01083-3.

45 **protein called cauxin:** Masao Miyazaki et al., "The Biological Function of Cauxin, a Major Urinary Protein of the Domestic Cat (*Felis catus*)," in *Chemical Signals in Vertebrates 11*, ed. Jane L. Hurst et al. (New York: Springer, 2008), 51–60.

45 **Felinine is synthesized:** Wouter H. Hendriks, Shane M. Rutherfurd, and Kay J. Rutherfurd, "Importance of Sulfate, Cysteine and Methionine as Precursors to Felinine Synthesis by Domestic Cats (*Felis catus*)," *Comparative Biochemistry and Physiology Part C: Toxicology & Pharmacology* 129, no. 3 (2001): 211–6.

46 **signals of fitness are common:** John W. S. Bradshaw, Rachel A. Casey, and Sarah L. Brown, "Communication," in *The Behaviour of the Domestic Cat*, 2nd ed. (Wallingford, UK: CABI, 2012), 91–112.

48 **when presented with three samples:** Miyabi Nakabayashi, Ryohei Yamaoka, and Yoshihiro Nakashima, "Do Fecal Odours Enable Domestic Cats (*Felis catus*) to Distinguish Familiarity of the Donors?" *Journal of Ethology* 30 (2012): 325–29, https://doi.org/10.1007/s10164-011-0321-x.

48 **Examining cat feces:** Ayami Futsuta et al., "LC-MS/MS Quantification of Felinine Metabolites in Tissues, Fluids, and Excretions from the Domestic Cat (*Felis catus*)," *Journal of Chromatography B* 1072 (2018): 94–9.

48 **derived from felinine:** Masao Miyazaki et al., "The Chemical Basis of Species, Sex, and Individual Recognition Using Feces in the Domestic Cat," *Journal of Chemical Ecology* 44 (2018): 364–73, https://doi.org/10.1007/s10886-018-0951-3.

49 **One survey found that 52 percent:** Colleen Wilson et al., "Owner Observations Regarding Cat Scratching Behavior: An Internet-Based Survey," *Journal of Feline Medicine and Surgery* 18, no. 10 (2016): 791–7.

49 **usually along routes:** Hilary Feldman, "Methods of Scent Marking in the Domestic Cat," *Canadian Journal of Zoology* 72, no. 6 (1994): 1093–9, https://doi.org/10.1139/z94-147.

53 **neuroanatomist Paul Broca:** Paul Broca, "Recherches sur les centres olfactifs," *Revue d'Anthropologie* 2 (1879): 385–455.

53 **scientists such as John McGann:** John P. McGann, "Poor Human Olfaction Is a 19th-Century Myth," *Science* 356, no. 6338 (2017): eaam7263.

53 **over a trillion different olfactory:** C. Bushdid et al., "Humans Can Discriminate More Than 1 Trillion Olfactory Stimuli," *Science* 343, no. 6177 (2014): 1370–2, https://doi.org/10.1126/science.1249168.

53 **One of the more entertaining:** Jess Porter et al., "Mechanisms of Scent-Tracking in Humans," *Nature Neuroscience* 10, no. 1 (2007): 27–9.

54 **A survey of four hundred people:** Ofer Perl et al., "Are Humans Constantly but Subconsciously Smelling Themselves?" *Philosophical Transactions of the Royal Society B* 375, no. 1800 (2020): 20190372.

55 **human interactive behavior:** Ida Frumin et al., "A Social Chemosignaling Function for Human Handshaking," *eLife* 4 (2015): e05154, https://doi.org/10.7554/eLife.05154.

55 **One small-scale study:** Nicola Courtney and Deborah L. Wells, "The Discrimination of Cat Odours by Humans," *Perception* 31, no. 4 (2002): 511–2.

56 **this particular smell:** Benjamin L. Hart and Mitzi G. Leedy, "Analysis of the Catnip Reaction: Mediation by Olfactory System, Not Vomeronasal Organ," *Behavioral and Neural Biology* 44, no. 1 (1985): 38–46.

57 **Geneticist Neil Todd:** Neil B. Todd, "Inheritance of the Catnip Response in Domestic Cats," *Journal of Heredity* 53, no. 2 (1962): 54–6, https://doi.org/10.1093/oxfordjournals.jhered.a107121.

57 **These include Tatarian honeysuckle:** Sebastiaan Bol et al., "Responsiveness of Cats (Felidae) to Silver Vine (*Actinidia polygama*), Tatarian Honeysuckle (*Lonicera tatarica*), Valerian (*Valeriana officinalis*) and Catnip (*Nepeta cataria*)," *BMC Veterinary Research* 13, no. 1 (2017): 1–16.

57 **Scientists looking at silver vine:** Reiko Uenoyama et al., "The Characteristic Response of Domestic Cats to Plant Iridoids Allows Them to Gain Chemical Defense Against Mosquitoes," *Science Advances* 7, no. 4 (2021): eabd9135.

57 **"Catnip: Its Raison d'Être":** Thomas Eisner, "Catnip: Its Raison d'Être," *Science* 146, no. 3649 (1964): 1318–20.

CHAPTER 3: YOU HAD ME AT MEOW

60 **diary entry by the Abbé Galiani:** Francis Steegmuller, *A Woman, a Man, and Two Kingdoms: The Story of Madame d'Épinay and the Abbé Galiani* (Princeton, NJ: Princeton University Press, 2014).

60 **Dupont de Nemours:** Described by Champfluery, "Cat Language," in *The Cat, Past and Present*, trans. Cashel Hoey (London: G. Bell, 1985).

61 ***Pussy and Her Language*:** Marvin R. Clark and Alphonse Leon Grimaldi, *Pussy and Her Language* (Fairford, UK: Echo Library, 2019).

62 **Mildred Moelk revolutionized:** Mildred Moelk, "Vocalizing in the House-Cat; a Phonetic and Functional Study," *American Journal of Psychology* 57, no. 2 (1944): 184–205.

63 **kittens have a distress call:** Ron Haskins, "A Causal Analysis of Kitten Vocalization: An Observational and Experimental Study," *Animal Behaviour* 27 (1979): 726–36.

64 **study by Wiebke Konerding:** Wiebke S. Konerding et al., "Female Cats, but Not Males, Adjust Responsiveness to Arousal in the Voice of Kittens," *BMC Evolutionary Biology* 16, no. 1 (2016): 1–9.

64 **each kitten develops its own:** Marina Scheumann et al., "Vocal Correlates of Sender-Identity and Arousal in the Isolation Calls of Domestic Kitten (*Felis silvestris catus*)," *Frontiers in Zoology* 9, no. 1 (2012): 1–14.

64 **these remain constant:** Robyn Hudson et al., "Stable Individual Differences in Separation Calls During Early Development in Cats and Mice," *Frontiers in Zoology* 12, suppl. 1 (2015): 1–12.

64 **writer Lafcadio Hearn:** Lafcadio Hearn, "Pathological," in *Kottō* (London: Macmillan and Co., Ltd., 1903).

65 **Researchers discovered this by:** Péter Szenczi et al., "Mother-Offspring Recognition in the Domestic Cat: Kittens Recognize Their Own Mother's Call," *Developmental Psychobiology* 58, no. 5 (2016): 568–77.

66 **describes the acoustics of the meow:** Nicholas Nicastro, "Perceptual and Acoustic Evidence for Species-Level Differences in Meow Vocalizations by Domestic Cats (*Felis catus*) and African Wild Cats (*Felis silvestris lybica*)," *Journal of Comparative Psychology* 118, no. 3 (2004): 287–96.

67 **more phonetic version:** Susanne Schötz, Joost van de Weijer, and Robert Eklund, "Melody Matters: An Acoustic Study of Domestic Cat Meows in Six Contexts and Four Mental States," *PeerJ Preprints* 7 (2019): e27926v1.

67 **Urban Dictionary's definition:** Urban Dictionary, s.v. "meow," last modified July 1, 2014, https://www.urbandictionary.com/define.php?term=Meow.

68 **an average frequency:** Katarina Michelsson, Helena Todd de Barra, and Oliver Michelson, "Sound Spectrographic Cry Analysis and Mothers' Perception of Their Infant's Crying," in *Focus on Nonverbal Communication Research*, ed. Finley R. Lewis (New York: Nova Science, 2007), 31–64.

68 **researchers, such as Schötz:** Susanne Schötz, Joost van de Weijer, and Robert Eklund, "Phonetic Methods in Cat Vocalization Studies: A Report from the Meowsic Project," in *Proceedings of the Fonetik*, vol. 2019 (Stockholm, 2019), 10–2.

68 **Joanna Dudek and coworkers:** Joanna Dudek et al., "Infant Cries Rattle Adult Cognition," *PLoS One* 11, no. 5 (2016): e0154283.

68 **take care of them like a baby:** The urgency with which adult humans respond to baby cries was investigated by Katherine S. Young et al. Testing adult men and women who weren't parents, they discovered that responses in the brain occurred earlier when hearing infant cries than

when hearing adult cries, suggesting the existence of a "caregiving instinct." Details of this can be found in Young et al., "Evidence for a Caregiving Instinct: Rapid Differentiation of Infant from Adult Vocalizations Using Magnetoencephalography," *Cerebral Cortex* 26, no. 3 (2016): 1309–21.

68 **the meows of domestic pet cats:** Nicholas Nicastro, "Perceptual and Acoustic Evidence for Species-Level Differences."

69 **acoustics of feral cat and pet cat:** Seong Yeon et al., "Differences Between Vocalization Evoked by Social Stimuli in Feral Cats and House Cats," *Behavioural Processes* 87, no. 2 (2011): 183–9.

69 **study by Fabiano de Oliveira Calleia:** Fabiano de Oliveira Calleia, Fábio Röhe, and Marcelo Gordo, "Hunting Strategy of the Margay (*Leopardus wiedii*) to Attract the Wild Pied Tamarin (*Saguinus bicolor*)," *Neotropical Primates* 16, no. 1 (2009): 32–4.

71 **use their barks:** Sophia Yin, "A New Perspective on Barking in Dogs (*Canis familaris*)," *Journal of Comparative Psychology* 116, no. 2 (2002): 189–93.

71 **Nicholas Nicastro recorded:** Nicholas Nicastro and Michael J. Owren, "Classification of Domestic Cat (*Felis catus*) Vocalizations by Naive and Experienced Human Listeners," *Journal of Comparative Psychology* 117, no. 1 (2003): 44–52.

72 **acoustic structure of barks:** Sophia Yin and Brenda McCowan, "Barking in Domestic Dogs: Context Specificity and Individual Identification," *Animal Behaviour* 68, no. 2 (2004): 343–55.

72 **later study by Sarah Ellis:** Sarah L. H. Ellis, Victoria Swindell, and Oliver H. P. Burman, "Human Classification of Context-Related Vocalizations Emitted by Familiar and Unfamiliar Domestic Cats: An Exploratory Study," *Anthrozoös* 28, no. 4 (2015): 625–34.

72 **higher scores on level of empathy:** Emanuela Prato-Previde et al., "What's in a Meow? A Study on Human Classification and Interpretation of Domestic Cat Vocalizations," *Animals* 10, no. 12 (2020): 2390.

73 **Tamás Faragó and his coworkers:** Tamás Faragó et al., "Humans Rely on the Same Rules to Assess Emotional Valence and Intensity in Conspecific and Dog Vocalizations," *Biology Letters* 10, no. 1 (2014): 20130926.

73 **concept first mooted by Darwin:** Charles Darwin, "Means of Expression in Animals," in *The Expression of the Emotions in Man and Animals* (New York: D. Appleton & Company, 1872), 83–114.

73 **One study found that meows:** M. A. Schnaider et al., "Cat Vocalization in Aversive and Pleasant Situations," *Journal of Veterinary Behavior* 55–56 (2022): 71–8.

73 **In a different study, Susanne Schötz:** Schötz, van de Weijer, and Eklund, "Melody Matters."

74 **cat-experienced listeners were better:** Susanne Schötz and Joost van de Weijer, "A Study of Human Perception of Intonation in Domestic Cat Meows," in *Social and Linguistic Speech Prosody: Proceedings of the 7th International Conference on Speech Prosody*, ed. Nick Campbell, Dafydd Gibbon, and Daniel Hirst (2014).

74 **Pascal Belin and coworkers:** Pascal Belin et al., "Human Cerebral Response to Animal Affective Vocalizations," *Proceedings of the Royal Society B: Biological Sciences* 275, no. 1634 (2008): 473–81.

75 **survey of people's impressions:** Christine E. Parsons et al., "Pawsitively Sad: Pet-Owners Are More Sensitive to Negative Emotion in Animal Distress Vocalizations," *Royal Society Open Science* 6, no. 8 (2019): 181555.

75 **Paul Gallico's amusing and charming:** Paul Gallico, *The Silent Miaow* (London: Pan Books Ltd., 1987).

75 **In one survey, 96 percent:** Victoria L. Voith and Peter L. Borchelt, "Social Behavior of Domestic Cats," in *Readings in Companion Animal Behavior*, ed. V. L. Voith and P. L. Borchelt (Trenton, NJ: Veterinary Learning Systems, 1996), 248–57.

76 **When they have been out:** Matilda Eriksson, Linda J. Keeling, and Therese Rehn, "Cats and Owners Interact More with Each Other After a Longer Duration of Separation," *PLoS One* 12, no. 10 (2017): e0185599.

76 **Research has shown that pet-directed:** Denis Burnham, Christine Kitamura, and Uté Vollmer-Conna, "What's New, Pussycat? On Talking to Babies and Animals," *Science* 296, no. 5572 (2002): 1435.

76 ***The Language Used in Talking to Domestic Animals*:** H. Carrington Bolton, "The Language Used in Talking to Domestic Animals," *American Anthropologist* 10, no. 3 (1897): 65–90.

77 **babies show a preference:** Tobias Grossmann et al., "The Developmental Origins of Voice Processing in the Human Brain," *Neuron* 65, no. 6 (2010): 852–8.

77 **mainly younger people:** Péter Pongrácz and Julianna Szulamit Szapu, "The Socio-Cognitive Relationship Between Cats and Humans—Companion Cats (*Felis catus*) as Their Owners See Them," *Applied Animal Behaviour Science* 207 (2018): 57–66.

77–78 **speech aimed deliberately:** Charlotte de Mouzon, Marine Gonthier, and Gérard Leboucher, "Discrimination of Cat-Directed Speech from Human-Directed Speech in a Population of Indoor Companion Cats (*Felis catus*)," *Animal Cognition* 26, no. 2 (2023): 611–9, https://doi.org/10.1007/s10071-022-01674-w.

78 **Cats' hearing range:** Rickye S. Heffner and Henry E. Heffner, "Hearing Range of the Domestic Cat," *Hearing Research* 19, no. 1 (1985): 85–8.

78 **Atsuko Saito and Kazutaka Shinozuka:** Atsuko Saito and Kazutaka Shinozuka, "Vocal Recognition of Owners by Domestic Cats (*Felis catus*)," *Animal Cognition* 16, no. 4 (2013): 685–90.

79 **cats' ability to distinguish:** Atsuko Saito et al., "Domestic Cats (*Felis catus*) Discriminate Their Names from Other Words," *Scientific Reports* 9, no. 5394 (2019): 1–8.

80 **We now know that it is controlled:** Dawn Frazer Sissom, D. A. Rice, and G. Peters, "How Cats Purr," *Journal of Zoology* 223, no. 1 (1991): 67–78.

81 **for thirty minutes:** Eriksson, Keeling, and Rehn, "Cats and Owners Interact More with Each Other."

81 **Karen Mccomb and her team:** Karen Mccomb et al., "The Cry Embedded Within the Purr," *Current Biology* 19, no. 13 (2009): R507–8.

CHAPTER 4: TALKATIVE TAILS AND EXPRESSIVE EARS

86 **Researchers have found that kangaroos:** Shawn M. O'Connor et al., "The Kangaroo's Tail Propels and Powers Pentapedal Locomotion," *Biology Letters* 10, no. 7 (2014): 20140381.

87 **structure of bone:** Emily Xu and Patricia M. Gray, "Evolutionary GEM: The Evolution of the Primate Prehensile Tail," *Western Undergraduate Research Journal: Health and Natural Sciences* 8, no. 1 (2017).

87 **California ground squirrel:** Matthew A. Barbour and Rulon W. Clark, "Ground Squirrel Tail-Flag Displays Alter Both Predatory Strike and Ambush Site Selection Behaviours of Rattlesnakes," *Proceedings of the Royal Society B: Biological Sciences* 279, no. 1743 (2012): 3827–33.

87 **Angelo Quaranta and coworkers:** A. Quaranta, M. Siniscalchi, and G. Vallortigara, "Asymmetric Tail-Wagging Responses by Dogs to Different Emotive Stimuli," *Current Biology* 17, no. 6 (2007): R199–201.

88 **Marcello Siniscalchi and co-researchers:** Marcello Siniscalchi et al., "Seeing Left- or Right-Asymmetric Tail Wagging Produces Different Emotional Responses in Dogs," *Current Biology* 23, no. 22 (2013): 2279–82.

88 **tend to keep their tails:** Daiana de Oliveira and Linda J. Keeling, "Routine Activities and Emotion in the Life of Dairy Cows: Integrating Body Language into an Affective State Framework," *PloS One* 13, no. 5 (2018): e0195674.

88 **cannibalistic tail biting:** Maya Wedin et al., "Early Indicators of Tail Biting Outbreaks in Pigs," *Applied Animal Behaviour Science* 208 (2018): 7–13.

89 **Dr. Amir Patel realized:** Amir Patel and Edward Boje, "On the Conical Motion and Aerodynamics of the Cheetah Tail," in *Robotics: Science and Systems Workshop on "Robotic Uses for Tails"* (Rome, 2015).

90 **described by E. W. Gudger:** Eugene Willis Gudger, "Does the Jaguar Use His Tail as a Lure in Fishing," *Journal of Mammalogy* 27, no. 1 (1946): 37–49.

91 **using their tails to communicate:** Sarah Louise Brown, "The Social Behaviour of Neutered Domestic Cats (*Felis catus*)" (PhD diss., University of Southampton, 1993).

93 **Following my early work:** John Bradshaw and Sarah Brown, "Social Behaviour of Cats," *Tijdschrift voor Diergeneeskunde* 177, no. 1 (1992): 54–6.

93 **tested the Tail Up behavior:** John Bradshaw and Charlotte Cameron-Beaumont, "The Signaling Repertoire of the Domestic Cat and Its Undomesticated Relatives," in *The Domestic Cat: The Biology of Its Behaviour,* 2nd ed., ed. Dennis C. Turner and Patrick Bateson (Cambridge, UK: Cambridge University Press, 2000), 67.

93 **Simona Cafazzo and Eugenia Natoli:** Simona Cafazzo and Eugenia Natoli, "The Social Function of Tail Up in the Domestic Cat (*Felis silvestris catus*)," *Behavioural Processes* 80, no. 1 (2009): 60–6.

94 **John Bradshaw looked:** John W. S. Bradshaw, "Sociality in Cats: A Comparative Review," *Journal of Veterinary Behavior* 11 (2016): 113–24.

95 **Penny Bernstein and Mickie Strack:** Penny L. Bernstein and Mickie Strack, "A Game of Cat and House: Spatial Patterns and Behavior of 14 Domestic Cats (*Felis catus*) in the Home," *Anthrozoös* 9, no. 1 (1996): 25–39.

95 **As part of my doctoral studies:** Brown, "The Social Behaviour of Neutered Domestic Cats."

96 **One of the most interesting:** This investigation of TU in wild felids and some discussion thereof can be found in Bradshaw and Cameron-Beaumont, "The Signaling Repertoire of the Domestic Cat."

96 **Comparative studies have:** Charlotte Cameron-Beaumont, "Visual and Tactile Communication in the Domestic Cat (*Felis silvestris catus*) and Undomesticated Small Felids" (PhD diss., University of Southampton, 1997).

98 **So how did the Tail Up:** The evolution of Tail Up is discussed in Cafazzo and Natoli, "The Social Function of Tail Up," and in Bradshaw and Cameron-Beaumont, "The Signaling Repertoire of the Domestic Cat."

98 **lions of the Serengeti:** George B. Schaller, *The Serengeti Lion: A Study of Predator-Prey Relations* (Chicago: University of Chicago Press, 1972).

100 **modern African wildcats:** David Macdonald et al., "African Wildcats in Saudi Arabia," *WildCRU Review* 42 (1996).

100 **Bateson and Turner suggested:** "Postscript: Questions and Some Answers," in *The Domestic Cat: The Biology of Its Behaviour,* 3rd ed., ed. Dennis C. Turner and Patrick Bateson (Cambridge: Cambridge University Press, 2014).

101 **Professor Alphonse Grimaldi:** Marvin R. Clark and Alphonse Leon Grimaldi, *Pussy and Her Language* (Fairford, UK: Echo Library, 2019).

102 **called the Facial Action Coding System:** The system has been updated over the years but its early development can be found in Paul Ekman and Wallace V. Friesen, "Measuring Facial Movement," *Environmental Psychology and Nonverbal Behavior* 1 (1976): 56–75, https://www.paulekman.com/wp-content/uploads/2013/07/Measuring-Facial-Movement.pdf.

102 **known as CatFACS:** Cátia Correia-Caeiro, Anne M. Burrows, and Bridget M. Waller, "Development and Application of CatFACS: Are Human Cat Adopters Influenced by Cat Facial Expressions?" *Applied Animal Behaviour Science* 189 (2017): 66–78.

105 **Bertrand Deputte and colleagues:** Bertrand L. Deputte et al., "Heads and Tails: An Analysis of Visual Signals in Cats, *Felis catus,*" *Animals* 11, no. 9 (2021): 2752.

107 **how people interpret:** Gabriella Tami and Anne Gallagher, "Description of the Behaviour of Domestic Dog (*Canis familiaris*) by Experienced and Inexperienced People," *Applied Animal Behaviour Science* 120, no. 3–4 (2009): 159–69.

108 **between cats and dogs living:** N. Feuerstein and Joseph Terkel, "Interrelationships of Dogs (*Canis familiaris*) and Cats (*Felis catus L.*) Living

Under the Same Roof," *Applied Animal Behaviour Science* 113, no. 1–3 (2008): 150–65.

CHAPTER 5: KEEPING IN TOUCH

112 **Robin Dunbar, exploring:** Robin I. M. Dunbar, "The Social Role of Touch in Humans and Primates: Behavioural Function and Neurobiological Mechanisms," *Neuroscience and Biobehavioral Reviews* 34, no. 2 (2010): 260–8.

113 **David Macdonald and Peter Apps:** David Macdonald et al., "Social Dynamics, Nursing Coalitions and Infanticide Among Farm Cats, *Felis catus*," *Advances in Ethology* (supplement to *Ethology*) 28 (1987): 1–64; David Macdonald, "The Pride of the Farmyard," *BBC Wildlife*, November 1991.

115 **large indoor colony:** Ruud van den Bos, "The Function of Allogrooming in Domestic Cats (*Felis silvestris catus*); a Study in a Group of Cats Living in Confinement," *Journal of Ethology* 16 (1998): 1–13.

116 **Allopreening in some species:** C. J. O. Harrison, "Allopreening as Agonistic Behaviour," *Behaviour* 24, no. 3/4 (1964): 161–209.

116 **nocturnal primate Garnett's bushbaby:** Jennie L. Christopher, "Grooming as an Agonistic Behavior in Garnett's Small-Eared Bushbaby (*Otolemur garnettii*)" (master's thesis, University of Southern Mississippi, 2017).

117 **occurs between adult dolphins:** Mai Sakai et al., "Flipper Rubbing Behaviors in Wild Bottlenose Dolphins (*Tursiops aduncus*)," *Marine Mammal Science* 22, no. 4 (2006): 966–78.

117 **Asian elephants, engaging:** Saki Yasui and Gen'ichi Idani, "Social Significance of Trunk Use in Captive Asian Elephants," *Ethology, Ecology & Evolution* 29, no. 4 (2017): 330–50, https://doi.org/10.1080/03949370.2016.1179684.

122 **Kimberly Barry and Sharon Crowell-Davis:** Kimberly J. Barry and Sharon L. Crowell-Davis, "Gender Differences in the Social Behavior of the Neutered Indoor-Only Domestic Cat," *Applied Animal Behaviour Science* 64, no. 3 (1999): 193–211.

123 **proposed for badgers:** Christina D. Buesching, P. Stopka, and D. W. Macdonald, "The Social Function of Allo-Marking in the European Badger (*Meles meles*)," *Behaviour* 140, no. 8/9 (2003): 965–80.

125 **this forward motion as "whiskers protractor":** More detailed descriptions and illustrations of whisker movements are available in the CatFACS manual: https://www.animalfacs.com/catfacs_new.

125 **Yngve Zotterman discovered:** Yngve Zotterman, "Touch, Pain and Tickling: An Electro-Physiological Investigation on Cutaneous Sensory Nerves," *Journal of Physiology* 95, no. 1 (1939): 1–28, https://doi.org/10.1113/jphysiol.1939.sp003707.

126 **creates the maximum response:** Rochelle Ackerley et al., "Human C-Tactile Afferents Are Tuned to the Temperature of a Skin-Stroking Caress," *Journal of Neuroscience* 34, no. 8 (2014): 2879–83.

126 **known as the insula:** Hakan Olausson et al., "Unmyelinated Tactile Afferents Signal Touch and Project to Insular Cortex," *Nature Neuroscience* 5, no. 9 (2002): 900–4.

127 **It seems to heighten:** Miranda Olff et al., "The Role of Oxytocin in Social Bonding, Stress Regulation and Mental Health: An Update on the Moderating Effects of Context and Interindividual Differences," *Psychoneuroendocrinology* 38, no. 9 (2013): 1883–94, https://doi.org/10.1016/j.psyneuen.2013.06.019; Simone G. Shamay-Tsoory and Ahmad Abu-Akel, "The Social Salience Hypothesis of Oxytocin," *Biological Psychiatry* 79, no. 3 (2016): 194–202, https://doi.org/10.1016/j.biopsych.2015.07.020.

127 **reduces sociality toward:** Annaliese K. Beery, "Antisocial Oxytocin: Complex Effects on Social Behavior," *Current Opinion in Behavioral Sciences* 6 (2015): 174–82, https://www.sciencedirect.com/science/article/pii/S2352154615001461.

129 **Claudia Mertens and Dennis Turner:** Claudia Mertens and Dennis C. Turner, "Experimental Analysis of Human-Cat Interactions During First Encounters," *Anthrozoös* 2, no. 2 (1988): 83–97.

129 **I carried out a study:** Sarah Louise Brown, "The Social Behaviour of Neutered Domestic Cats (*Felis catus*)" (PhD diss., University of Southampton, 1993).

131 **"Tripping over the Cat":** Bruce R. Moore and Susan Stuttard, "Dr. Guthrie and *Felis domesticus* or: Tripping over the Cat," *Science* 205, no. 4410 (1979): 1031–3.

131 **Edwin Guthrie and George Horton:** E. R. Guthrie and G. P. Horton, *Cats in a Puzzle Box* (New York: Rinehart, 1946).

132 **other subtle differences:** Claudia Mertens, "Human-Cat Interactions in the Home Setting," *Anthrozoös* 4, no. 4 (1991): 214–31.

132 **rubbing does not increase:** Matilda Eriksson, Linda J. Keeling, and Therese Rehn, "Cats and Owners Interact More with Each Other After a Longer Duration of Separation," *PLoS One* 12, no. 10 (2017): e0185599. And also Matilda Eriksson, "The Effect of Time Left Alone on Cat Behaviour" (master's thesis, University of Uppsala, 2015).

132 **Dogs have been shown:** Therese Rehn and Linda J. Keeling, "The Effect of Time Left Alone at Home on Dog Welfare," *Applied Animal Behaviour Science* 129, no. 2–4 (2011): 129–35.

132 **high rate around dinnertime:** John W. S. Bradshaw and Sarah E. Cook, "Patterns of Pet Cat Behaviour at Feeding Occasions," *Applied Animal Behaviour Science* 47, no. 1–2 (1996): 61–74.

133 **Edwards and co-researchers:** Claudia Edwards et al., "Experimental Evaluation of Attachment Behaviors in Owned Cats," *Journal of Veterinary Behavior* 2, no. 4 (2007): 119–25.

134 **tactile part was important:** Therese Rehn et al., "Dogs' Endocrine and Behavioural Responses at Reunion Are Affected by How the Human Initiates Contact," *Physiology & Behavior* 124 (2014): 45–53.

134 **A study by Nadine Gourkow:** N. Gourkow, S. C. Hamon, and C. J. C. Phillips, "Effect of Gentle Stroking and Vocalization on Behaviour,

Mucosal Immunity and Upper Respiratory Disease in Anxious Shelter Cats," *Preventive Veterinary Medicine* 117, no. 1 (2014): 266–75.

134 **stroking was more effective:** Sita Liu et al., "The Effects of the Frequency and Method of Gentling on the Behavior of Cats in Shelters," *Journal of Veterinary Behavior* 39 (2020): 47–56.

135 **lead their owner:** Penny Bernstein, "The Human-Cat Relationship," in *The Welfare of Cats*, ed. Irene Rochlitz (Dordrecht, Netherlands: Springer, 2007), 47–89.

135 **Sarah Ellis and co-researchers:** Sarah L. H. Ellis et al., "The Influence of Body Region, Handler Familiarity and Order of Region Handled on the Domestic Cat's Response to Being Stroked," *Applied Animal Behaviour Science* 173 (2015): 60–7.

136 **Dairy cows, for example:** Claudia Schmied et al., "Stroking of Different Body Regions by a Human: Effects on Behaviour and Heart Rate of Dairy Cows," *Applied Animal Behaviour Science* 109, no. 1 (2008): 25–38.

136 **looked into stroking:** Chantal Triscoli et al., "Touch Between Romantic Partners: Being Stroked Is More Pleasant Than Stroking and Decelerates Heart Rate," *Physiology & Behavior* 177 (2017): 169–75.

136 **Researchers looking at women:** Elizabeth A. Johnson et al., "Exploring Women's Oxytocin Responses to Interactions with Their Pet Cats," *PeerJ* 9 (2021): e12393.

137 **inferior frontal gyrus:** Ai Kobayashi et al., "The Effects of Touching and Stroking a Cat on the Inferior Frontal Gyrus in People," *Anthrozoös* 30, no. 3 (2017): 473–86, https://doi.org/10.1080/08927936.2017.1335115.

138 **cat owners from Brazil:** Daniela Ramos and Daniel S. Mills, "Human Directed Aggression in Brazilian Domestic Cats: Owner Reported Prevalence, Contexts and Risk Factors," *Journal of Feline Medicine and Surgery* 11, no. 10 (2009): 835–41, https://doi.org/10.1016/j.jfms.2009.04.006.

138 **The CT afferent fibers:** Chantal Triscoli, Rochelle Ackerley, and Uta Sailer, "Touch Satiety: Differential Effects of Stroking Velocity on Liking and Wanting Touch over Repetitions," *PLoS One* 9, no. 11 (2014): e113425.

138 **One interesting study:** Cátia Correia-Caeiro, Anne M. Burrows, and Bridget M. Waller, "Development and Application of CatFACS: Are Human Cat Adopters Influenced by Cat Facial Expressions?" *Applied Animal Behaviour Science* 189 (2017): 66–78.

139 **British vet James Herriot:** James Herriot, *James Herriot's Cat Stories*, 2nd ed. (New York: St. Martin's Press, 2015).

CHAPTER 6: SEEING EYE TO EYE

144 **Experiments by Phyllis Chesler:** Phyllis Chesler, "Maternal Influence in Learning by Observation in Kittens," *Science* 166, no. 3907 (1969): 901–3, https://doi.org/10.1126/science.166.3907.901.

145 **study of observational learning:** E. Roy John et al., "Observation Learning in Cats," *Science* 159, no. 3822 (1968): 1489–91, https://doi.org/10.1126/science.159.3822.1489.

146 **The theory of object permanence:** Jean Piaget, *The Construction of Reality in the Child*, trans. Margaret Cook (Oxford, UK: Routledge, 2013).

146 **shown to succeed:** Sonia Goulet, François Y. Doré, and Robert Rousseau, "Object Permanence and Working Memory in Cats (*Felis catus*)," *Journal of Experimental Psychology: Animal Behavior Processes* 20, no. 4 (1994): 347–65, https://doi.org/10.1037/0097-7403.20.4.347.

147 **first thirty seconds after an object:** Sylvain Fiset and François Y. Doré, "Duration of Cats' (*Felis catus*) Working Memory for Disappearing Objects," *Animal Cognition* 9, no. 1 (2006): 62–70, https://doi.org/10.1007/s10071-005-0005-4.

148 **When researcher Hitomi Chijiiwa:** Hitomi Chijiiwa et al., "Dogs and Cats Prioritize Human Action: Choosing a Now-Empty Instead of a Still-Baited Container," *Animal Cognition* 24, no. 1 (2021): 65–73.

150 **Researcher Jane Dards:** Jane L. Dards, "The Behaviour of Dockyard Cats: Interactions of Adult Males," *Applied Animal Ethology* 10, no. 1–2 (1983): 133–53.

151 **Deborah Goodwin and John Bradshaw:** Unpublished data in: John Bradshaw and Charlotte Cameron-Beaumont, "The Signaling Repertoire of the Domestic Cat and Its Undomesticated Relatives," in *The Domestic Cat: The Biology of Its Behaviour,* 2nd ed., ed. Dennis C. Turner and Patrick Bateson (Cambridge: Cambridge University Press, 2000).

151 **to record eye contact:** Deborah Goodwin and John W. S. Bradshaw, "Gaze and Mutual Gaze: Its Importance in Cat/Human and Cat/Cat Interactions," Conference Proceedings of the International Society for Anthrozoology (Boston, 1997).

152 **famous sociologist Georg Simmel:** Georg Simmel, "Sociology of the Senses: Visual Interaction," in *Introduction to the Science of Sociology,* eds. E. R. Park and E. W. Burgess (Chicago: University of Chicago Press, 1921), 356–61.

153 **we prefer mutual gazes:** Nicola Binetti et al., "Pupil Dilation as an Index of Preferred Mutual Gaze Duration," *Royal Society Open Science* 3, no. 7 (2016): 160086, http://dx.doi.org/10.1098/rsos.160086.

153 **One small study showed:** Deborah Goodwin and John W. S. Bradshaw, "Regulation of Interactions Between Cats and Humans by Gaze and Mutual Gaze," Abstracts from International Society for Anthrozoology Conference (Prague, 1998).

153 **how pet dogs and cats:** Marine Grandgeorge et al., "Visual Attention Patterns Differ in Dog vs. Cat Interactions with Children with Typical Development or Autism Spectrum Disorders," *Frontiers in Psychology* 11 (2020): 2047.

155 **Ádám Miklósi and coworkers:** Ádám Miklósi et al., "A Comparative Study of the Use of Visual Communicative Signals in Interactions Between Dogs (*Canis familiaris*) and Humans and Cats (*Felis catus*) and Humans," *Journal of Comparative Psychology* 119, no. 2 (2005): 179–86, https://doi.org/10.1037/0735-7036.119.2.179.

155 **Lingna Zhang and co-researchers:** Lingna Zhang et al., "Feline Communication Strategies When Presented with an Unsolvable Task: The Attentional State of the Person Matters," *Animal Cognition* 24, no. 5 (2021): 1109–19.

156 **study by Lea Hudson:** Lea M. Hudson, "Comparison of Canine and Feline Gazing Behavior" (Honors College thesis, Oregon State University, 2018), https://ir.library.oregonstate.edu/concern/honors_college_theses/m900p083f.

157 **Péter Pongrácz and coworkers:** Péter Pongrácz, Julianna Szulamit Szapu, and Tamás Faragó, "Cats (*Felis silvestris catus*) Read Human Gaze for Referential Information," *Intelligence* 74 (2019): 43–52.

158 **found in dogs:** Tibor Tauzin et al., "The Order of Ostensive and Referential Signals Affects Dogs' Responsiveness When Interacting with a Human," *Animal Cognition* 18, no. 4 (2015): 975–9, https://doi.org/10.1007/s10071-015-0857-1.

160 **Ádám Miklósi and co-researchers also discovered:** Miklósi et al., "A Comparative Study of the Use of Visual Communicative Signals."

160 **cats were at least as good at the task as:** Ádám Miklosi and Krisztina Soproni, "A Comparative Analysis of Animals' Understanding of the Human Pointing Gesture," *Animal Cognition* 9 (2006): 81–93.

160 **explored human-cat pointing:** Péter Pongrácz and Julianna Szulamit Szapu, "The Socio-Cognitive Relationship Between Cats and Humans—Companion Cats (*Felis catus*) as Their Owners See Them," *Applied Animal Behaviour Science* 207 (2018): 57–66.

162 **A survey of cat owners in Hungary:** "Moggies Remain a Mystery to Many, Suggests Survey," Cats Protection, https://www.cats.org.uk/mediacentre/pressreleases/behaviour-survey.

162 **Tasmin Humphrey and co-researchers:** Tasmin Humphrey et al., "The Role of Cat Eye Narrowing Movements in Cat-Human Communication," *Scientific Reports* 10, no. 1 (2020): 16503.

163 **A second study by Humphrey:** Tasmin Humphrey et al., "Slow Blink Eye Closure in Shelter Cats Is Related to Quicker Adoption," *Animals* 10, no. 12 (2020): 2256.

165 **Guillaume-Benjamin-Amand Duchenne de Boulogne:** Guillaume-Benjamin Duchenne de Boulogne, *The Mechanism of Human Facial Expression,* trans. R. Andrew Cuthbertson (Cambridge: Cambridge University Press, 1990).

165 **imitate this eye-wrinkling:** Sarah D. Gunnery, Judith A. Hall, and Mollie A. Ruben, "The Deliberate Duchenne Smile: Individual Differences in Expressive Control," *Journal of Nonverbal Behavior* 37, no. 1 (2013): 29–41.

CHAPTER 7: THE PERSONALITY PUZZLE

167 **children's book *Six-Dinner Sid*:** Inga Moore, *Six-Dinner Sid* (New York: Aladdin, 2004).

169 **in the 1970s by Jerome Kagan:** Roger G. Kuo, "Psychologist Finds Shyness Inherited, but Not Permanent," *Harvard Crimson,* March 4, 1991,

https://www.thecrimson.com/article/1991/3/4/psychologist-finds-shyness-inherited-but-not/.

169 **often referred to as the Big Five:** This topic has been studied by many researchers over the years—a summary can be found in Christopher J. Soto and Joshua J. Jackson, "Five-Factor Model of Personality," in *Oxford Bibliographies in Psychology*, ed. Dana S. Dunn (New York: Oxford University Press, 2020).

172 **Melanie Dammhahn and coworkers:** Melanie Dammhahn et al., "Of City and Village Mice: Behavioural Adjustments of Striped Field Mice to Urban Environments," *Scientific Reports* 10, no. 1 (2020): 13056.

173 **Bolder male cats:** Eugenia Natoli et al., "Bold Attitude Makes Male Urban Feral Domestic Cats More Vulnerable to Feline Immunodeficiency Virus," *Neuroscience and Biobehavioral Reviews* 29, no. 1 (2005): 151–7.

174 **Alert, Sociable, and Equable:** Julie Feaver, Michael Mendl, and Patrick Bateson, "A Method for Rating the Individual Distinctiveness of Domestic Cats," *Animal Behaviour* 34, no. 4 (1986): 1016–25.

175 **Known as the Feline Five:** Carla Litchfield et al., "The 'Feline Five': An Exploration of Personality in Pet Cats (*Felis catus*)," *PLoS One* 12, no. 8 (2017): e0183455.

179 **in the 1980s, a groundbreaking discovery:** Eileen B. Karsh and Dennis C. Turner, "The Human-Cat Relationship," in *The Domestic Cat: The Biology of Its Behaviour*, ed. Dennis C. Turner and Patrick G. Bateson (Cambridge: Cambridge University Press, 1988), 159–77.

181 **Researching cat paternity:** Sandra McCune, "The Impact of Paternity and Early Socialisation on the Development of Cats' Behaviour to People and Novel Objects," *Applied Animal Behaviour Science* 45, no. 1–2 (1995): 109–24.

184 **Using DNA samples:** Data is from Figures 1 and 2 in the paper: Ludovic Say, Dominique Pontier, and Eugenia Natoli, "High Variation in Multiple Paternity of Domestic Cats (*Felis catus L.*) in Relation to Environmental Conditions," *Proceedings of the Royal Society B: Biological Sciences* 266, no. 1433 (1999): 2071–4.

185 **Certain traits appear to be established:** Sarah E. Lowe and John W. S. Bradshaw, "Ontogeny of Individuality in the Domestic Cat in the Home Environment," *Animal Behaviour* 61, no. 1 (2001): 231–7.

187 **Dr. Rush Shippen Huidekoper's:** Rush Shippen Huidekoper, *The Cat, a Guide to the Classification and Varieties of Cats and a Short Treatise upon Their Care, Diseases, and Treatment* (New York: D. Appleton, 1895).

187 **Mikel Delgado and colleagues:** Mikel M. Delgado, Jacqueline D. Munera, and Gretchen M. Reevy, "Human Perceptions of Coat Color as an Indicator of Domestic Cat Personality," *Anthrozoös* 25, no. 4 (2012): 427–40, https://doi.org/10.2752/175303712X13479798785779.

187 **In a later survey of cat owners:** Mónica Teresa González-Ramírez and René Landero-Hernández, "Cat Coat Color, Personality Traits and the Cat-Owner Relationship Scale: A Study with Cat Owners in Mexico," *Animals* 12, no. 8 (2022): 1030, https://doi.org/10.3390/ani12081030.

188 **harder to read the emotions:** Haylie D. Jones and Christian L. Hart, "Black Cat Bias: Prevalence and Predictors," *Psychological Reports* 123, no. 4 (2020): 1198–206.

188 **adoption of black cats:** Lori R. Kogan, Regina Schoenfeld-Tacher, and Peter W. Hellyer, "Cats in Animal Shelters: Exploring the Common Perception That Black Cats Take Longer to Adopt," *Open Veterinary Science Journal* 7, no. 1 (2013).

188 **Korat and Devon Rex:** Milla Salonen et al., "Breed Differences of Heritable Behaviour Traits in Cats," *Scientific Reports* 9, no. 1 (2019): 7949.

189 **oxytocin receptor (OXTR) gene:** Minori Arahori et al., "The Oxytocin Receptor Gene (OXTR) Polymorphism in Cats (*Felis catus*) Is Associated with 'Roughness' Assessed by Owners," *Journal of Veterinary Behavior* 11 (2016): 109–12.

189 **In a somewhat unsettling:** Michael M. Roy and Nicholas J. S. Christenfeld, "Do Dogs Resemble Their Owners?" *Psychological Science* 15, no. 5 (2004): 361–3.

189 **A small but intriguing study:** Lawrence Weinstein and Ralph Alexander, "College Students and Their Cats," *College Student Journal* 44, no. 3 (2010): 626–8.

190 **how different personalities interact:** Kurt Kotrschal et al., "Human and Cat Personalities: Building the Bond from Both Sides," in *The Domestic Cat: The Biology of Its Behaviour*, 3rd ed., ed. Dennis C. Turner and Patrick Bateson (Cambridge, UK: Cambridge University Press, 2014), 113–29.

190 **anxious owner–cat relationship:** Lauren R. Finka et al., "Owner Personality and the Wellbeing of Their Cats Share Parallels with the Parent-Child Relationship," *PloS One* 14, no. 2 (2019): e0211862.

191 **Despite being more intense:** Manuela Wedl et al., "Factors Influencing the Temporal Patterns of Dyadic Behaviours and Interactions Between Domestic Cats and Their Owners," *Behavioural Processes* 86, no. 1 (2011): 58–67.

191 **Interactions initiated by the cat:** Dennis C. Turner, "The Ethology of the Human-Cat Relationship," *Schweizer Archiv fur Tierheilkunde* 133, no. 2 (1991): 63–70.

191 **Kurt Kotrschal and co-writers:** Kotrschal et al., "Human and Cat Personalities."

CHAPTER 8: THE PLEASURE OF THEIR COMPANY

196 **Claudia Mertens and Dennis Turner in 1988:** Claudia Mertens and Dennis C. Turner, "Experimental Analysis of Human-Cat Interactions During First Encounters," *Anthrozoös* 2, no. 2 (1988): 83–97.

198 **encounters in their own homes:** Claudia Mertens, "Human-Cat Interactions in the Home Setting," *Anthrozoös* 4, no. 4 (1991): 214–31.

199 **With respect to interaction reciprocity:** Dennis C. Turner, "The Mechanics of Social Interactions Between Cats and Their Owners," *Frontiers in Veterinary Science* 8 (2021): 292.

199 **mooted for rhesus monkeys:** Robert A. Hinde, "On Describing Relationships," *Journal of Child Psychology and Psychiatry* 17, no. 1 (1976): 1–19.

200 **study by Manuela Wedl:** Manuela Wedl et al., "Factors Influencing the Temporal Patterns of Dyadic Behaviours and Interactions Between Domestic Cats and Their Owners," *Behavioural Processes* 86, no. 1 (2011): 58–67.

200 **As Dennis Turner points out:** This is discussed by Dennis C. Turner in his summary paper "The Mechanics of Social Interactions Between Cats and Their Owners."

203 **other cats may simply tolerate:** Daniela Ramos et al., "Are Cats (*Felis catus*) from Multi-Cat Households More Stressed? Evidence from Assessment of Fecal Glucocorticoid Metabolite Analysis," *Physiology & Behavior* 122 (2013): 72–5.

204 **Camilla Haywood and co-researchers:** Camilla Haywood et al., "Providing Humans with Practical, Best Practice Handling Guidelines During Human-Cat Interactions Increases Cats' Affiliative Behaviour and Reduces Aggression and Signs of Conflict," *Frontiers in Veterinary Science* 8 (2021): 835.

205 **The prize for this:** W. L. Alden, "Postal Cats," in *Domestic Explosives and Other Sixth Column Fancies* (New York: Lovell, Adam, Wesson & Co., 1877), 192–4, https://archive.org/details/domesticexplosi00aldegoog/page/n6/mode/2up.

206 **More recently, consideration:** Regina M. Bures, "Integrating Pets into the Family Life Cycle," in *Well-Being Over the Life Course*, ed. Regina M. Bures and Nancy R. Gee (New York: Springer, 2021), 11–23.

207 **In a fascinating study:** Esther M. C. Bouma, Marsha L. Reijgwart, and Arie Dijkstra, "Family Member, Best Friend, Child or 'Just' a Pet, Owners' Relationship Perceptions and Consequences for Their Cats," *International Journal of Environmental Research and Public Health* 19, no. 1 (2021): 193.

209 **Cat Owner Relationship Scale:** Tiffani J. Howell et al., "Development of the Cat-Owner Relationship Scale (CORS)," *Behavioural Processes* 141, no. 3 (2017): 305–15.

209 **Dog Owner Relationship Scale:** Fleur Dwyer, Pauleen C. Bennett, and Grahame J. Coleman, "Development of the Monash Dog Owner Relationship Scale (MDORS)," *Anthrozoös* 19, no. 3 (2006): 243–56.

209 **social exchange theory:** Richard M. Emerson, "Social Exchange Theory," *Annual Review of Sociology* 2 (1976): 335–62.

210 **experience sampling method:** Mayke Janssens et al., "The Pet-Effect in Daily Life: An Experience Sampling Study on Emotional Wellbeing in Pet Owners," *Anthrozoös* 33, no. 4 (2020): 579–88.

211 **high on the Neuroticism dimension:** Gretchen M. Reevy and Mikel M. Delgado, "The Relationship Between Neuroticism Facets, Conscientiousness, and Human Attachment to Pet Cats," *Anthrozoös* 33, no. 3 (2020): 387–400, https://doi.org/10.1080/08927936.2020.1746527.

211 **cats make frequent physical contact:** Pim Martens, Marie-José Enders-Slegers, and Jessica K. Walker, "The Emotional Lives of Companion Animals: Attachment and Subjective Claims by Owners of Cats and Dogs," *Anthrozoös* 29, no. 1 (2016): 73–88.

212 **Strange Situation Test (SST):** Mary Ainsworth's work can be explored in more detail in the following sources: Mary D. S. Ainsworth et al., *Strange Situation Procedure (SSP)*, APA PsycNet (1978), https://doi.org/10.1037/t28248-000; Mary Ainsworth et al., *Patterns of Attachment: A Psychological Study of the Strange Situation* (London: Psychology Press, 2015).

212 **Some studies have indicated:** József Topál et al., "Attachment Behavior in Dogs (*Canis familiaris*): A New Application of Ainsworth's (1969) Strange Situation Test," *Journal of Comparative Psychology* 112, no. 3 (1998): 219–29.

212 **However it has also been observed:** Elyssa Payne, Pauleen C. Bennett, and Paul D. McGreevy, "Current Perspectives on Attachment and Bonding in the Dog-Human Dyad," *Psychology Research and Behavior Management* 8 (2015): 71–9.

212 **Two of these found:** Claudia Edwards et al., "Experimental Evaluation of Attachment Behaviors in Owned Cats," *Journal of Veterinary Behavior* 2, no. 4 (2007): 119–25; Kristyn R. Vitale, Alexandra C. Behnke, and Monique A. R. Udell, "Attachment Bonds Between Domestic Cats and Humans," *Current Biology* 29, no. 18 (2019): R864–5.

213 **cats did not regard their owners:** Alice Potter and Daniel S. Mills, "Domestic Cats (*Felis silvestris catus*) Do Not Show Signs of Secure Attachment to Their Owners," *PLoS One* 10, no. 9 (2015): e0135109.

213 **Mauro Ines and co-researchers:** Mauro Ines, Claire Ricci-Bonot, and Daniel S. Mills, "My Cat and Me—a Study of Cat Owner Perceptions of Their Bond and Relationship," *Animals* 11, no. 6 (2021): 1601.

215 **We attribute to our pets:** Martens, Enders-Slegers, and Walker, "The Emotional Lives of Companion Animals."

217 **over 70 percent reported:** Ashley L. Elzerman et al., "Conflict and Affiliative Behavior Frequency Between Cats in Multi-Cat Households: A Survey-Based Study," *Journal of Feline Medicine and Surgery* 22, no. 8 (2020): 705–17.

218 **form part of the Facial Action Coding System:** The system has been updated over the years but its early development can be found in Paul Ekman and Wallace V. Friesen, "Measuring Facial Movement," *Environmental Psychology and Nonverbal Behavior* 1 (1976): 56–5, https://www.paulekman.com/wp-content/uploads/2013/07/Measuring-Facial-Movement.pdf.

218 **Dog Facial Action Coding System:** Bridget M. Waller et al., "Paedomorphic Facial Expressions Give Dogs a Selective Advantage," *PLoS One* 8, no. 12 (2013): e82686.

219 **recorded the facial movements of cats:** Cátia Correia-Caeiro, Anne M. Burrows, and Bridget M. Waller, "Development and Application of

CatFACS: Are Human Cat Adopters Influenced by Cat Facial Expressions?" *Applied Animal Behaviour Science* 189 (2017): 66–78.

220 **Lauren Dawson and colleagues:** Lauren Dawson et al., "Humans Can Identify Cats' Affective States from Subtle Facial Expressions," *Animal Welfare* 28, no. 4 (2019): 519–31.

220 **Moriah Galvan and Jennifer Vonk:** Moriah Galvan and Jennifer Vonk, "Man's Other Best Friend: Domestic Cats (*F. silvestris catus*) and Their Discrimination of Human Emotion Cues," *Animal Cognition* 19, no. 1 (2016): 193–205.

221 **A later study by Angelo Quaranta:** This study also tested cats on their ability to recognize matching visual and vocal sounds of other cats. Angelo Quaranta et al., "Emotion Recognition in Cats," *Animals* 10, no. 7 (2020): 1107.

222 **One study put this to the test:** Kristyn R. Vitale Shreve, Lindsay R. Mehrkam, and Monique A. R. Udell, "Social Interaction, Food, Scent or Toys? A Formal Assessment of Domestic Pet and Shelter Cat (*Felis silvestris catus*) Preferences," *Behavioural Processes* 141, no. 3 (2017): 322–8.

EPILOGUE: THE ADAPTABLE CAT

225 ***They Walked Beside Me*:** Katharine L. Simms, *They Walked Beside Me* (London: Hutchison and Co., 1954), 99.

228 **Who needs opposable thumbs anyway?:** After hearing about Smudge's latest escape trick, the cat flap company went into action once more and made us a special add-on piece for the cat flap to prevent Smudge from hooking it open. So far it has held her at bay, but I'm sure she's secretly plotting her next escape.

INDEX

Note: Italicized page numbers indicate material in illustrations.

INDEX

ABOUT THE AUTHOR

Sarah Brown gained her PhD on the social behavior of neutered domestic cats while working at the Anthrozoology Institute at the University of Southampton in the United Kingdom. She has since worked as an independent cat behavior counselor, as a consultant for the cat-toy industry, and has conducted research for and worked with several UK animal charities. She authored *The Cat: A Natural and Cultural History*, which has been published in three languages; cowrote *The Behaviour of the Domestic Cat*, 2nd edition; and contributed to *The Domestic Cat: The Biology of Its Behaviour*, 3rd edition. Sarah lives in London with her family; her dog, Reggie; and her cats, Bootsy and Smudge.